SBC State Machine for Model-Based Systems Engineering

-- Toward a Unified View of the System --

William S. Chao • Shuh-Ping Sun

Structure-Behavior Coalescence

Systems Architecture = Systems Structure + Systems Behavior

CONTENTS

PREFACE

Systems Modeling Language (SysML) is a general modeling language for model-based systems engineering (MBSE) applications. The SysML specification defines a set of language concepts that is used to model the (static) structure and (dynamic) behavior of a system. The SysML concepts include (1) an abstract syntax that defines the language concepts and is described by a metamodel, and (2) a concrete syntax, or notation, that defines how the language concepts are represented and is described by a user model.

Since SysML is a multi-diagram approach, there are always some inconsistencies between different diagrams in the user model. To ensure and check the consistency, a metamodel that defines the abstract syntax of a modeling language needs to provide a unified semantic framework for defining consistency rules to impose constraints on the structure (i.e., blocks) or behavior (i.e., activities) constructs. It is hoped that through this unified semantic framework, each diagram in the user model can be projected as a view of the metamodel.

Unfortunately, most current SysML metamodels do not have the ability to project each diagrams in the user model as a view of the metamodel. In this book, we developed SBC State Machine (SSM) as a metamodel for SysML. In SBC State Machine, each diagram in the user model will be projected as a view of the metamodel. Therefore, we claim that SBC State Machine genuinely provides a unified semantic framework to ensure model consistency for SysML.

ABOUT THE AUTHORS

Dr. William S. Chao is the CEO & founder of SBC Architecture International®. SBC (Structure-Behavior Coalescence) architecture is a systems architecture which demands the integration of systems structure and systems behavior of a system. SBC architecture applies to hardware architecture, software architecture, enterprise architecture, knowledge architecture and thinking architecture. The core theme of SBC architecture is: Architecture = Structure + Behavior.

William S. Chao received his bachelor degree (1976) in telecommunication engineering and master degree (1981) in information engineering, both from the National Chiao-Tung University, Taiwan. From 1976 till 1983, he worked as an engineer at Chung-Hwa Telecommunication Company, Taiwan.

William S. Chao received his master degree (1985) in information science and Ph.D. degree (1988) in information science, both from the University of Alabama at Birmingham, USA. From 1988 till 1991, he worked as a computer scientist at GE Research and Development Center, Schenectady, New York, USA.

Dr. William S. Chao has been teaching at National Sun Yat-Sen University, Taiwan from 1992 till 2019 and now serves as the president of Association of Enterprise Architects, Taiwan Chapter. His research covers: systems architecture, hardware architecture, software architecture, enterprise architecture, knowledge architecture and thinking architecture.

Dr. Shuh-Ping Sun was born in Taiwan. He received a Ph.D. degree in Mechanical Engineering from the Auburn University, AL, USA. He was an associate professor during 1995-2008, and has been a professor since Aug. 2008, in the Department of Biomedical Engineering at I-Shou University, Taiwan. He is a professor in the Department of Digital Media Design and the director of System Architecture research center at I-Shou University. He conducts research in Systems Architecture (SBC Architecture) and Creative Media Design.

PART I: MODEL-BASED SYSTEMS ENGINEERING

Chapter 1: Introduction to Model-Based Systems Engineering

As a general application of modeling to support system requirements, design, analysis, verification, and validation tasks beginning in the conceptual design phase and continuing throughout development and later life cycle phases, model-based systems engineering (MBSE) aims to promote systems engineering activities that have traditionally been performed using the document-based approach and result in enhanced specification and design quality, reuse of system specification and design artifacts, as well as communication between development teams.

In this chapter, we first discuss what model-based systems engineering is. Then, we compare the document-based approach with the model-based approach for systems engineering.

1-1 What is Model-Based Systems Engineering?

Model-based systems engineering (MBSE) is a systems engineering method that concentrates on creating and exploiting a model as the primary means of information exchange between systems engineers during in the system development process, as shown in Figure 1-1.

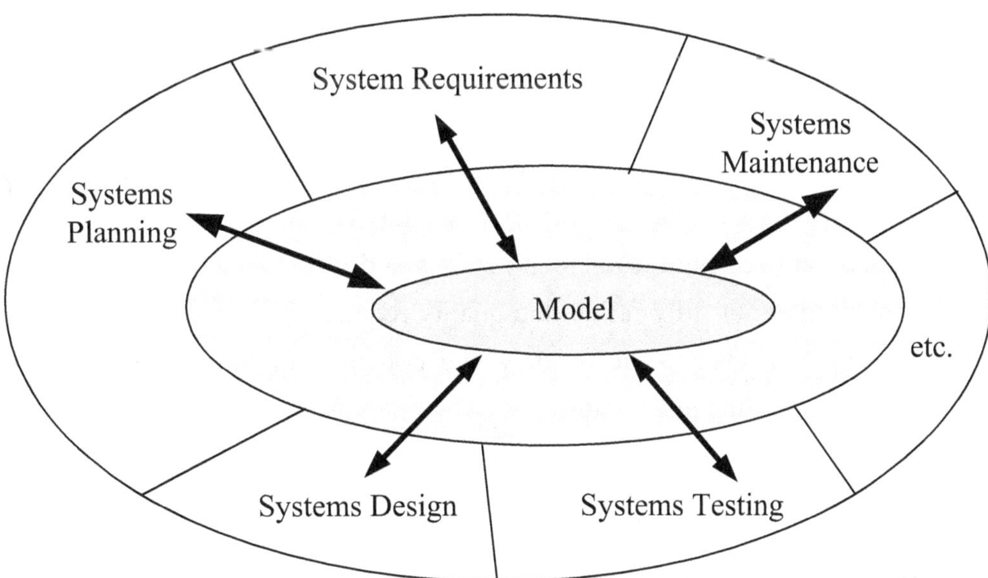

Figure 1-1. Concept of Model-Based Systems Engineering

In Figure 1-1, we see that systems planners, systems analysts, systems designers, systems testers, and systems maintainers all use this model to communicate and exchange information during work.

1-2 Document-Based Approach Versus Model-Based Approach

Document-based systems engineering is the traditional approach, while model-based systems engineering is the advanced approach.

1-2-1 Document-Based Systems Engineering Approach

Traditionally, large-scale projects use document-based systems engineering approaches to perform the system engineering activities. This approach is characterized by generating text descriptions and design documents, using hard copy or the electronic file format is then exchanged between customers, users, developers and testers. The system requirements and design information are expressed in these documents and drawings.

The focus of document-based systems engineering is on controlling documents and ensuring documents and the drawings are effective, complete and consistent, and the developed system meets documentation.

The document-based systems engineering may be strict, but there are some basic limitations. The relationship between completeness, consistency and requirements, design, engineering analysis and test information are difficult to evaluate because the information is distributed in multiple documents. This makes it difficult to understand specific aspects of the system and difficult to implement the necessary traceability and change impact assessments. In turn, this leads to poor synchronization between high-level requirements and lower-level design. The status of the document may not fully reflect the quality required by the system and design. These limitations can lead to inefficiency and potential quality issues that usually show up during integration and testing.

1-2-2 Model-Based Systems Engineering Approach

Model-based systems engineering integrates system requirements, designs, analyzes and validates models to solve various aspects of the system in a cohesive manner, rather than a disparate collection of individual models.

Model-based systems engineering provides an opportunity to indicate the many limitations of the document-based systems engineering by providing more rigorous means to capture and integrate system requirements, analyze, design and verify information, and promote maintenance, evaluation and communication of this information throughout the life cycle of the system.

System engineering aims to combine various parts of the system into a cohesive whole. Model-based systems engineering must support this basic focus of systems engineering. Especially pay much attention to that the role of the coherent model is to provide an integrated framework for models created by other engineering disciplines, including software, hardware and testing.

Chapter 2: Core Theme of Model-Based Systems Engineering

In this chapter, we first discuss the output of the model-based systems engineering activities. Then we discuss the core theme of model-based systems engineering.

2-1 Output of the MBSE Activities

The output of the model-based systems engineering activities is a coherent model of the system (i.e., system model) as shown in Figure 2-1, where the emphasis is placed on evolving and refining this coherent model using model-based methods and tools.

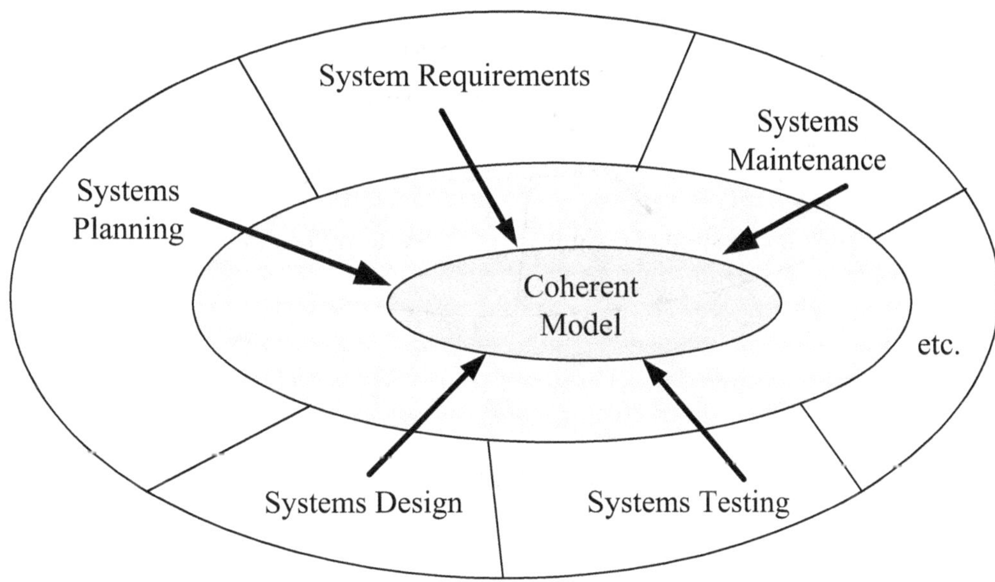

Figure 2-1. Output of the MBSE Activities

In Figure 2-1, we see that systems planners, systems analysts, systems designers, systems testers, and systems maintainers all use this coherent model to communicate and exchange information during work.

2-2 Core Theme of MBSE

The core theme of MBSE is a consistent model, i.e., systems modeling language (SysML) as shown in Figure 2-2, of the system's (static) structure and (dynamic) behavior, with an emphasis on using model-based methods and tools to develop and improve the model.

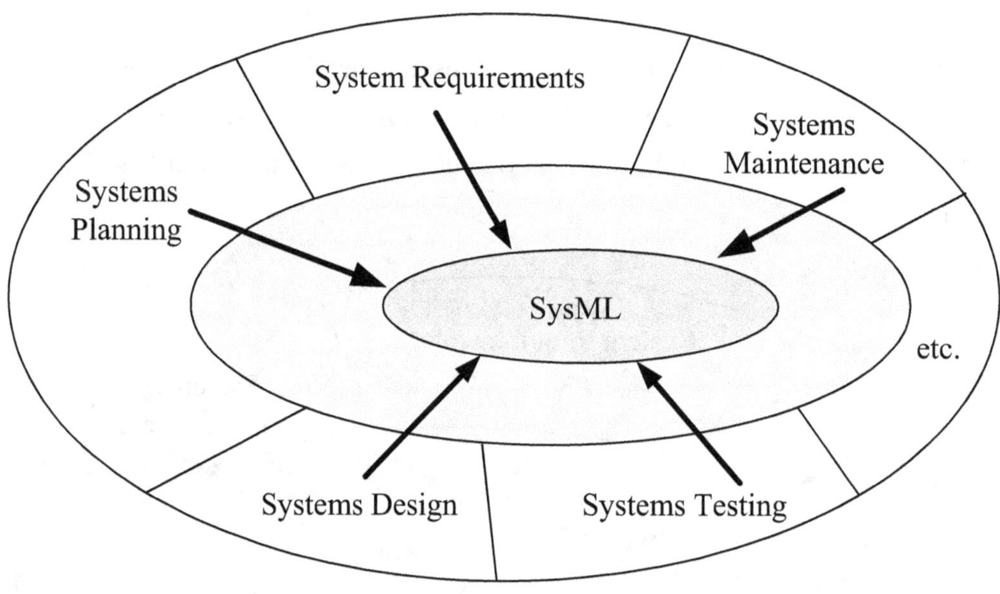

Figure 2-1. Core Theme of MBSE

In Figure 2-2, we see that systems planners, systems analysts, systems designers, systems testers, and systems maintainers all use this SysML to communicate and exchange information during work.

Chapter 3: Inconsistency Problems of Model-Based Systems Engineering

The core theme of the MBSE is a consistent model, i.e., systems modeling language (SysML), of the system's (static) structure and (dynamic) behavior, with an emphasis on using model-based methods and tools to develop and improve the model. However, since SysML is a multi-diagram approach, there are always unavoidable inconsistencies between different diagrams in the SysML specification of a system.

3-1 Systems Modeling Language

Systems Modeling Language (SysML) [Dell13, Frie14] is a general modeling language for model-based systems engineering (MBSE) applications [Dori16, INCO04]. The SysML specification defines a set of language concepts that is used to model the static structure and dynamic behavior of a system. The SysML concepts include (a) an abstract syntax that defines the language concepts and is represented and described by a metamodel, and (b) a concrete syntax, or notation, that defines how the language concepts are represented and is described by a user model (i.e., systems model) [Weil08].

SysML user model utilizes at least two types of diagrams, as shown in Figure 3-1, to represent different views of the system under consideration. (A) Structure type: emphasizes the static structure of the system using blocks, attributes, operations and relationships. This structure type includes block definition diagram, internal block diagram, package diagram, parametric diagram, etc. (B) Behavior type: emphasizes the dynamic behavior of the system by showing collaborations among blocks and changes to the internal states of blocks. This behavior type includes use case diagram, state machine, activity diagram, sequence diagrams, etc.

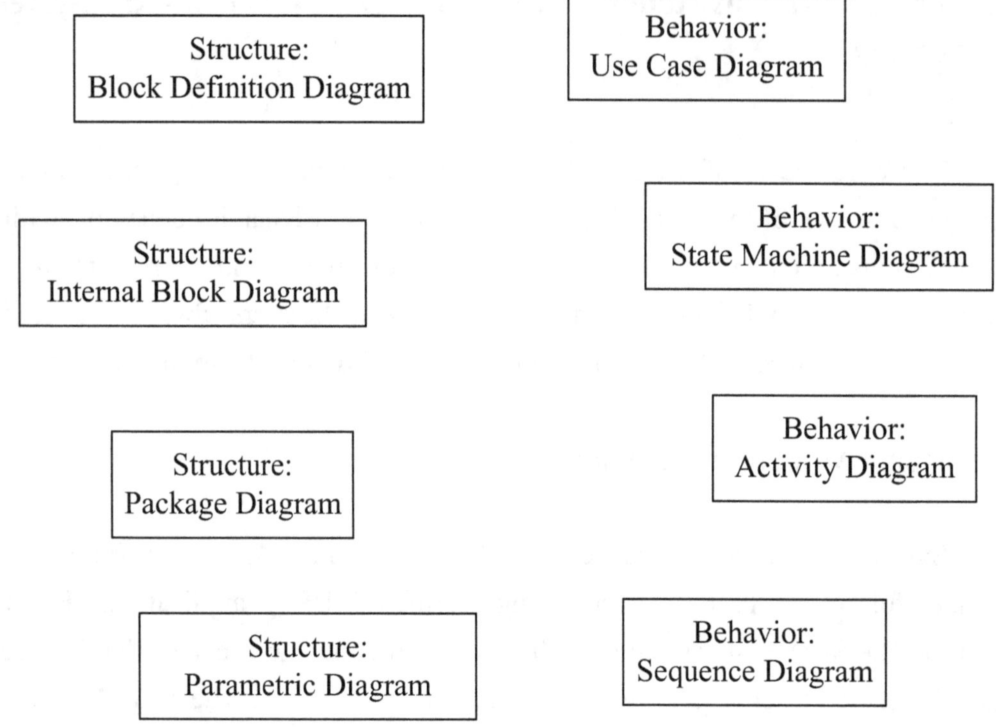

Figure 3-1. Multiple Diagrams of SysML User Model

3-2 Inconsistency Problems of SysML

Since SysML user model is a multi-diagram approach, there are inevitable inconsistencies [Alla15, Dori95, Dori02, Dori16, Enge02, Malg06, Pele00, Przi16] between those different diagrams, as shown in Figure 3-2. In a multi-diagram environment, comprehending a system and the way it operates and changes over time requires concurrent reference to the various diagrams and the creation of abstract associations that link them. These multiple diagrams are separated. Rather than being built into the method, the mental burden of integrating the various diagrams is placed on the shoulder of the developers who need to deal with a system that is complex in itself, and they are mentally overloaded without any reason. Technical solutions that involve sophisticated tools can reduce manual consistency maintenance, but do not address the central issues of excessive mental burden.

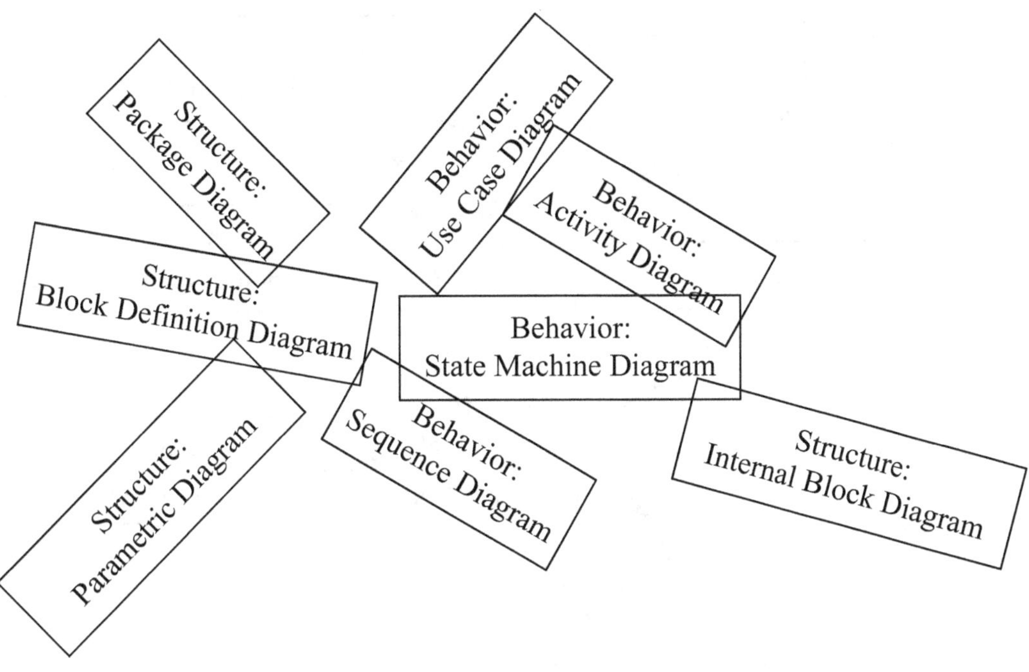

Figure 3-2. Inevitable Inconsistencies between Different Diagrams

In the SysML multi-diagram environment, the straightforward intuition of thinking concurrently about the structure and behavior is seriously hindered by separating the structure and behavior diagrams. These multiple diagrams are dissociated and always contradictory with each other, which become the major cause for the SysML inconsistency problems [Alla15, Bash16, Dori16].

3-3 Metamodel of SysML

To ensure and check the consistency, we always need to create a kernel model for SysML. This kernel model is the metamodel of SysML. All SysML structure views such as block definition diagram, internal block diagram and SysML behavior views such as use case diagram, state machine, activity diagram, sequence diagram, can be projected from this SysML metamodel, as shown in Figure 3-3.

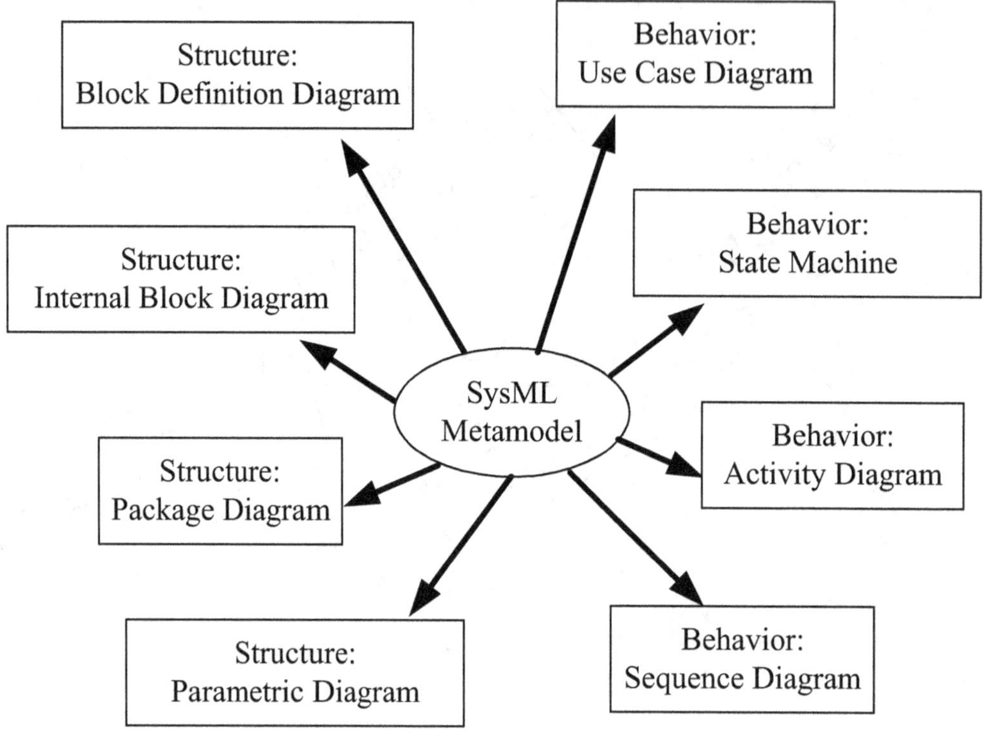

Figure 3-3. Metamodel of SysML

In SysML, an important usage of metamodels is to ensure model consistency between different diagrams in the user model. However, most existing SysML metamodels lack the capability to project each diagram in the user model as a view of the metamodel. In the next chapter, we will detail the shortcomings of the current SysML metamodel approaches.

3-4 Deficiency of Current SysML Metamodels

In SysML, a metamodel [Lale08] is used to describe the concepts in the language, their characteristics, and interrelationships. This is sometimes called the abstract syntax of the language, and is distinct from the concrete syntax that specifies the user model for the language. A significant usage of the metamodel is to ensure model consistency between different diagrams in the user model. However, most current SysML metamodels lack the capability to project each diagram in the user model as a view of the metamodel because they cannot provide a unified semantic framework.

The Object Management Group (OMG) defines a language for representing metamodels, called Meta Object Facility (MOF) that is used to define UML, SysML and other metamodels. Several mechanisms are used in MOF, such as Object Constraint Language (OCL), Foundational UML (fUML), The Action Language for Foundational UML (Alf), Process Specification Language (PSL), to name a few.

The Object Constraint Language (OCL) is a precise text language that provides constraint on the structure (i.e., objects) to ensure consistency of the user model [Przi16]. However, not every diagram in the user model can be projected as a view of the OCL metamodel because the OCL fails to provide a unified semantic framework. Therefore, the OCL metamodel can only ensure part of (not all) user model consistency.

The Foundational UML is a subset of the standard UML for which a standard execution constraint language, PSL, is used to define the semantics of the execution model [OMG13a]. Although fUML provides constraint on the behavior (i.e., activities) to make the model executable, it fails to integrate the structural constructs with the behavioral constructs. Not being able to provide a unified semantic framework, the Foundational UML can not project every diagram in the user model as a view of the fUML metamodel.

The Action Language for Foundational UML (Alf) is a complementary specification to Foundational UML [OMG13b]. The key use of Alf is to act as the notation for specifying executable behaviors in SysML, for example, methods for object operations, the behavior of an object, or transition effects on state machines. Like fUML, Alf also fails to provide a unified semantic framework to integrate the structural constructs with the behavioral constructs. Therefore, the Alf is not able to project every diagram in the user model as a view of the Alf metamodel.

Chapter 4: Why Using SBC State Machine for Model-Based Systems Engineering?

Systems structure and systems behavior are the two most prominent views of a system, integrating the systems structure and systems behavior is apparently the best way to achieve an integrated whole of a system. If we are not able to integrate the systems structure and systems behavior, then there is no way that we are able to integrate the whole system. Structure-behavior coalescence (SBC) provides an elegant way to integrate the systems structure and systems behavior of a system. In other words, SBC facilitates an integrated whole of a system.

4-1 Structure-Behavior Coalescence Means to Integrate the Systems Structure and Systems Behavior

All things that strike us as something independent are essentially parts of a system. We usually call the parts of a system its blocks. Blocks are sometimes labeled as components, objects, parts, entities, building blocks and non-aggregated systems [Chao14a, Chao14b, Chao14c, Chao16a].

Systems structure, specified by blocks, their operations and their composition, refers to the type of connection between the blocks of a system as shown in Figure 4-1.

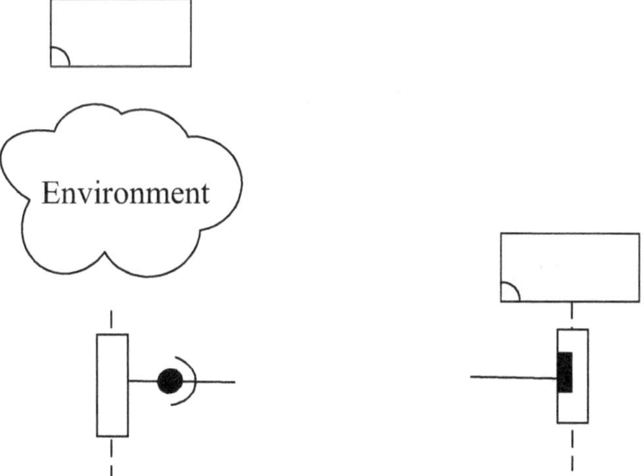

Figure 4-1. Systems Structure

Systems behavior, specified by the flow of interactions between and among the blocks and environment, refers to the interconnectivities a system in conjunction with its environment as shown in Figure 4-2.

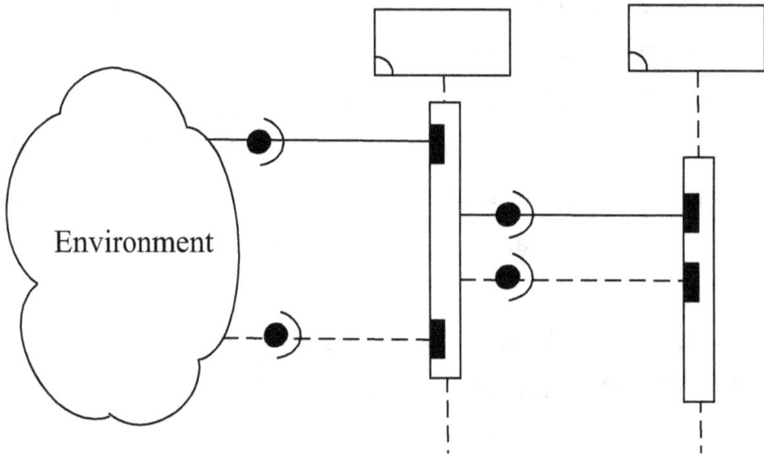

Figure 4-2. Systems Behavior

Systems structure and systems behavior are the two most prominent views of a system, integrating the systems structure and systems behavior is apparently the best way to achieve an integrated whole of a system.

If we are not able to integrate the systems structure and systems behavior, then there is no way that we are able to integrate the whole system.

Structure-behavior coalescence (SBC) [Chao15a, Lin19] provides an elegant way to integrate the systems structure and systems behavior of a system. In other words, SBC facilitates an integrated whole of a system.

4-2 Kernel Concept of Structure-Behavior Coalescence

The kernel concept of structure-behavior coalescence is: "Systems Architecture = Systems Structure + Systems Behavior." That is, the systems structure will lead to the systems behavior as shown in Figure 4-3.

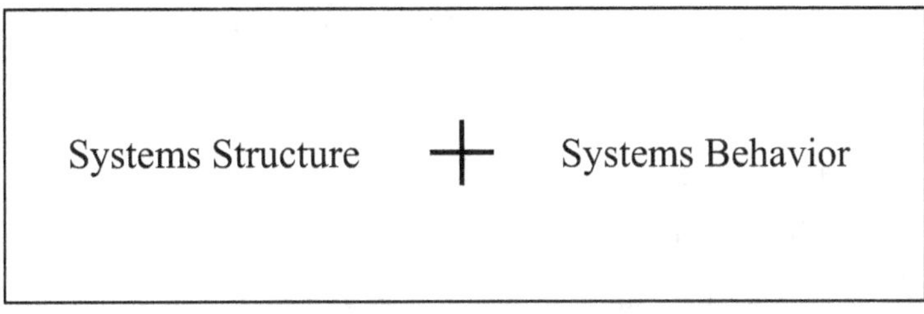

Figure 4-3. Kernel Concept of Structure-Behavior Coalescence

One systems structure will draw forth one systems behavior. That is, the systems behavior is attached to or built on the systems structure in the SBC approach.

In other words, the systems behavior can not exist alone; it must be loaded on the systems structure just like a cargo is loaded on a ship as shown in Figure 4-4.

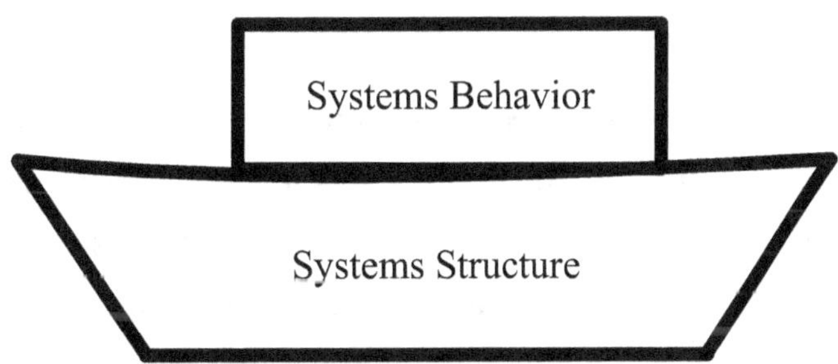

Figure 4-4. Systems Behavior Must be Loaded on the Systems Structure

4-3 Model Multiplicity Versus Model Singularity

In general, a system is extremely intricate that it contains several aspects, or "views", such as data, structure, function, behavior, and so on. There are two approaches to model these different views. The multi-model approach, also known as the model multiplicity approach, respectively utilizes a distinct model for each view. In a multi-model environment, comprehending a system and the way it operates and changes over time requires simultaneous reference to the different models and the construction of abstract associations that connect them. These multiple models are heterogeneous and separated and therefore inconsistent with each other, which become the primary reason for the model multiplicity problems of the multi-model approach

On the contrary, the single model approach also known as the model singularity approach, instead of choosing many separated models, will use only one integrated model. All structure, data, function and behavior views can be derived from this single model.

The use of multiple models to delineate a system from different views is the major reason for the model multiplicity problems. Being able to think about a system in one single integrated model, the model singularity approach truly avoids the model multiplicity problems.

4-4 SBC State Machine for Model-Based Systems Engineering

In order to overcome the shortcomings of the current SysML metamodel approaches, we need to develop a unified semantic framework that is able to integrate the structural constructs with the behavioral constructs.

SBC state machine is a labelled transition system (LTS) [Miln89, Miln99] which provides a single diagram to integrate structural and behavioral constructs in the MBSE modeling construction. Therefore, the model consistency will be fully ensured in the SBC state machine modeling construction. Using SBC State Machine as a metamodel for SysML, each diagram in the user model can be projected as a view of the metamodel, as shown in Figure 4-5.

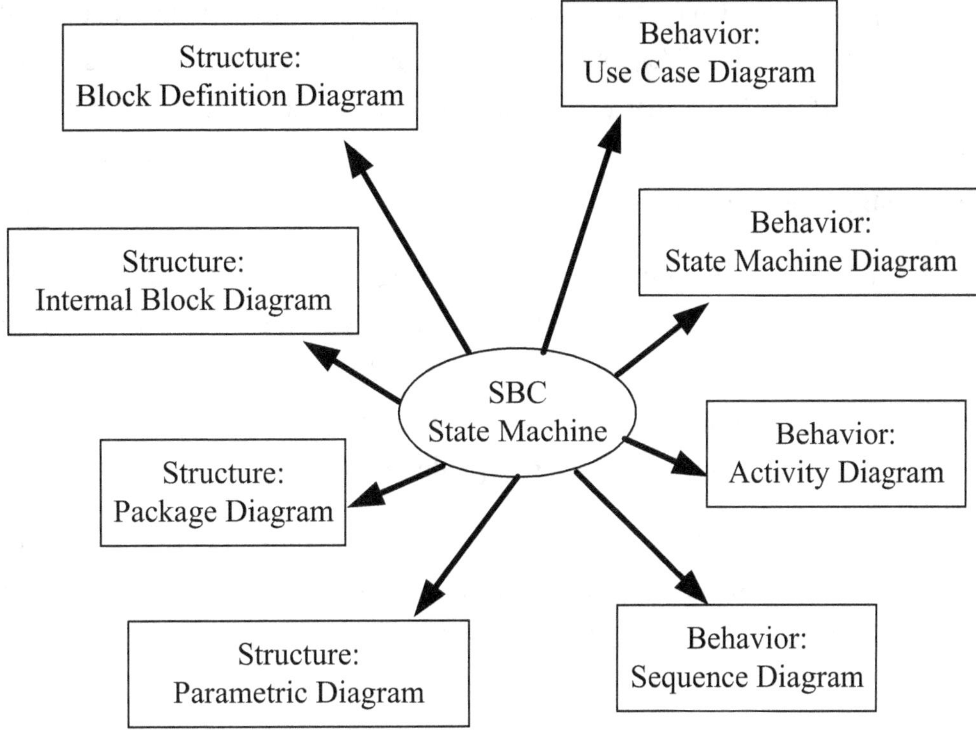

Figure 4-5. SBC State Machine as the Metamodel of SysML

4-5 Origin of SBC State Machine

SBC state machine originates from SBC process algebra. There are generalized SBC process algebra and specialized SBC process algebra.

4-5-1 Generalized SBC Process Algebra

Generalized SBC process algebra (G-SBC-PA) evolved from CCS (Calculus of Communicating Systems) [Miln89, Miln99].

CCS is a general process algebra language for the study of communication and concurrency. Like CCS, generalized SBC process algebra is also a general process algebra language for the study of communication and concurrency.

4-5-2 Specialized SBC Process Algebra

Channel-based single-queue SBC process algebra (C-S-SBC-PA) [Chao15d, Chao15e, Chao15g], channel-based multi-queue SBC process algebra (C-M-SBC-PA) [Chao15d, Chao15f, Chao15h], channel-based infinite-queue SBC process algebra (C-

I-SBC-PA) [Chao15b, Chao15c, Chao15d], operation-based single-queue SBC process algebra (O-S-SBC-PA), operation-based multi-queue SBC process algebra (O-M-SBC-PA) [Chao15d, Chao15f, Chao15h] and operation-based infinite-queue SBC process algebra (O-I-SBC-PA) [Chao15b, Chao15c, Chao15d] are the six specialized SBC process algebras.

Channel-based single-queue SBC process algebra, channel-based multi-queue SBC process algebra, channel-based infinite-queue SBC process algebra, operation-based single-queue SBC process algebra, operation-based multi-queue SBC process algebra and operation-based infinite-queue SBC process algebra all evolved from CCS (Calculus of Communicating Systems) [Miln89, Miln99].

CCS is a general process algebra language for the study of concurrent systems. Unlike CCS, six specialized SBC process algebras are only applicable to systems model [Burd10, Maie09, Chao16b, Chao16c, Chao16d, Chao16e, Chao16f, Chao16g, Craw15, Chec99, Dam06, O'Rou03, Putm00, Rayn09, Roza11, Toga08].

PART II: SBC STATE MACHINE

Chapter 5: Channel-Based Value-Passing Interactions

In this chapter, we first introduce channels and channel-based value-passing interactions. We then introduce the formal description of channel-based communication ports, channel-based value-passing actions and channel-based value-passing interactions.

5-1 Channels

A block can **provide** multiple channels for agent communication. We use a component channel diagram (CChD), as shown in Figure 5-1, to display the channels of all blocks in the system.

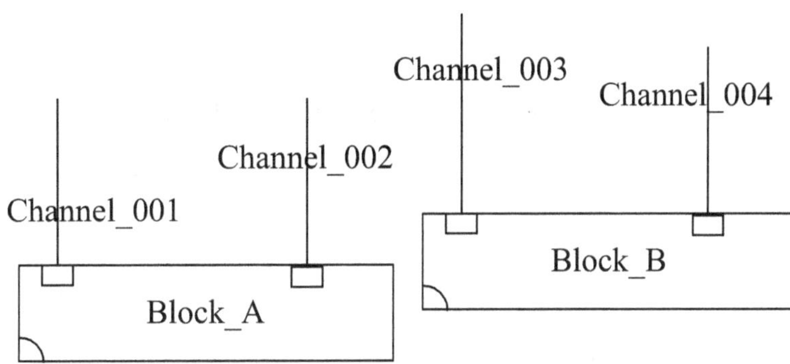

Figure 5-1. Component Channel Diagram

A channel may contain several input parameters (e.g. i_1, i_2) and output parameters (e.g. o_1, o_2), as shown in Figure 5-2.

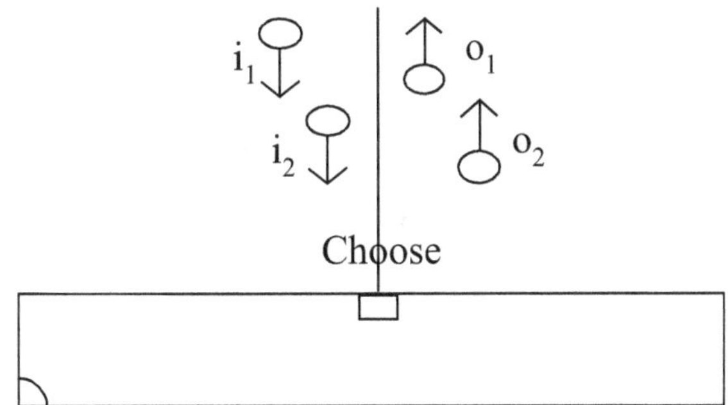

Figure 5-2. A Channel Contains Several Input/Output Parameters

A channel signature is used to completely describe a channel. The signature for a channel is a combination of its name along with parameters as follows:

<channel name> (<parameter list>)

The parameters in the parameter list represent the inputs or outputs of the channel. Each parameter in the list is displayed with the following format:

<direction> <parameter name> : <parameter type>

Parameter direction may be in, out, or inout. We formally describe the "channel signature" as a relation $K \subseteq \Lambda \times \Theta$ where Λ is a set of "channel names" and Θ is a set of "parameter lists".

DEFINITION (CHANNEL SIGNATURE) A Channel Signature CS = (Λ, Θ, K) consists of

. a finite set Λ of "channel names",
. a finite set Θ of "parameter lists",
. a relation $K \subseteq \Lambda \times \Theta$, and $(ch, p) \in K$.

5-2 Component Channel Diagram

We formally describe the "component channel diagram" as a relation $CChD \subseteq \Lambda \times \Theta \times \Gamma$ where Λ is a set of "channel names" and Θ is a set of "parameter lists" and Γ is a set of "blocks".

DEFINITION (COMPONENT CHANNEL DIAGRAM) A Component Channel Diagram CChD = $(\Lambda, \Theta, \Gamma, CChD)$ consists of

. a finite set Λ of "channel names",
. a finite set Θ of "parameter lists",
- a finite set Γ of "blocks",
- a relation $CChD \subseteq \Lambda \times \Theta \times \Gamma$, and $(ch, p, b) \in CChD$.

5-3 Channel-Based Value-Passing Interactions

An interaction represents an indivisible and instantaneous handshake or communication between two agents. In the channel-based value-passing approach as shown in Figure 5-3, the caller agent (either external environment's actor or block) communicates with the callee agent (block) through the channel interaction.. Figure 5-3 also depicts that the "Choose (In p, q; Out r, s)" channel is **required** by the caller agent and is **provided** by the callee agent.

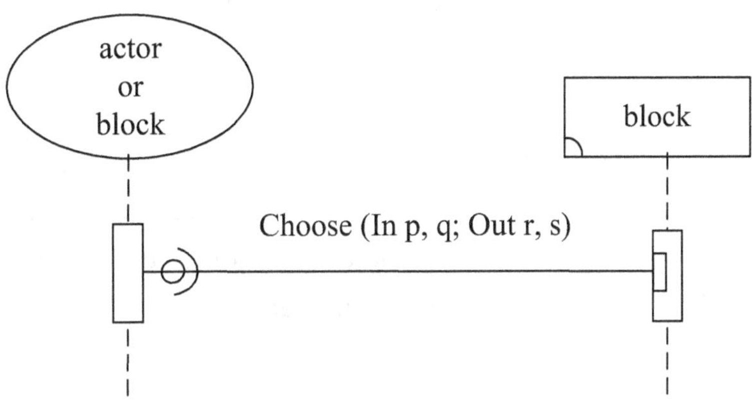

Figure 5-3. Channel-Based Value-Passed Interaction

The caller agent owns the "calling port" of the interaction. In this case, the calling port is "$\overline{\text{Choose (In p, q; Out r, s)}}$" and its conduct is to assist the caller agent to output a value (send a message) to each of the "p" and "q" variables (of the "Choose" channel), and input a value (receive a message) from each of the "r" and "s" variables (of the "Choose" channel), as shown in Figure 5-4.

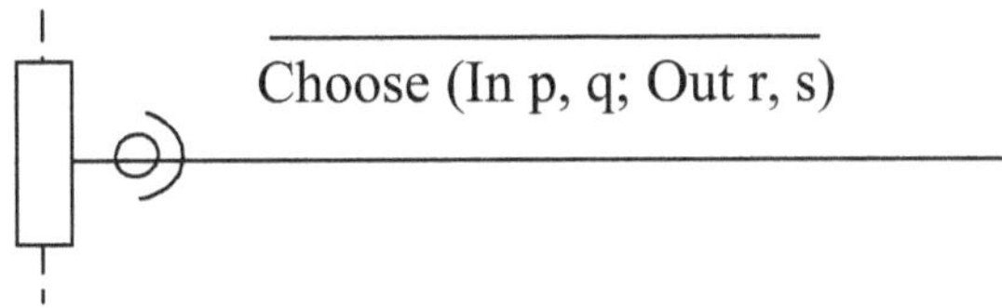

Figure 5-4. Calling Port

The caller agent together with the "calling port" is named the "calling action" as shown in Figure 5-5.

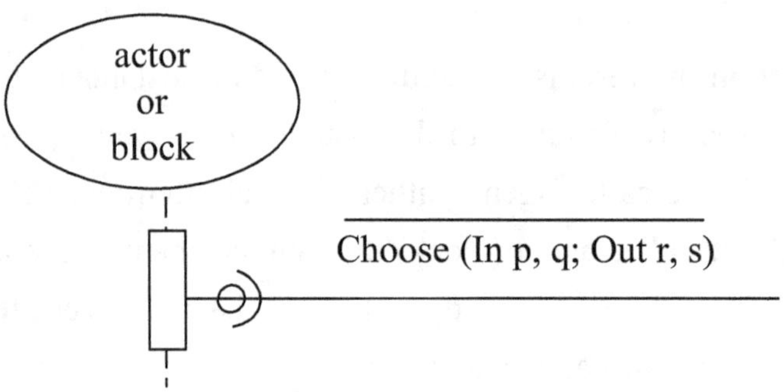

Figure 5-5. Calling Action

The callee agent owns the "called port" of the interaction. In this case, the called port is "Choose (In p, q; Out r, s)" and its conduct is to assist the callee agent to input a value (receive a message) for each of the "p" and "q" variables (of the "Choose" channel), and output a value (send a message) for each of the "r" and "s" variables (of the "Choose" channel), as shown in Figure 5-6.

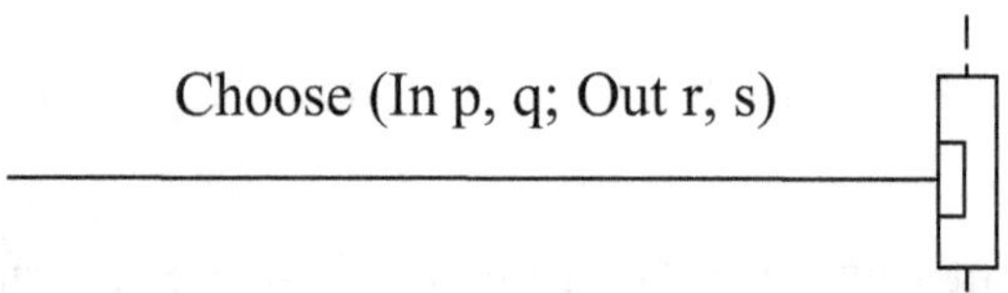

Figure 5-6. Called Port

The callee agent together with the "called port" is named the "called action" as shown in Figure 5-7.

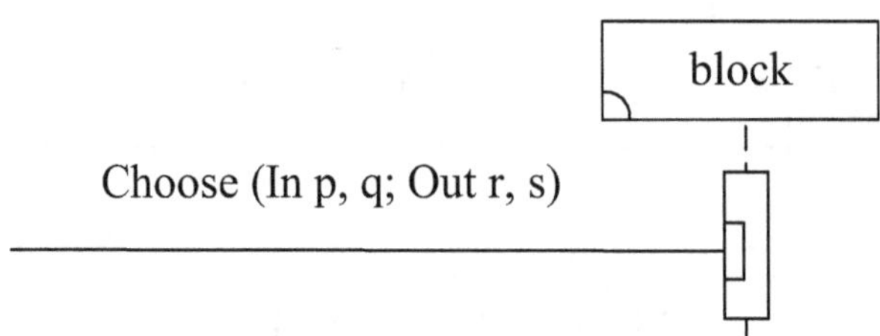

Figure 5-7. Called Action

Also we can draw the "channel-based value-passing interaction" diagram as shown in Figure 5-8.

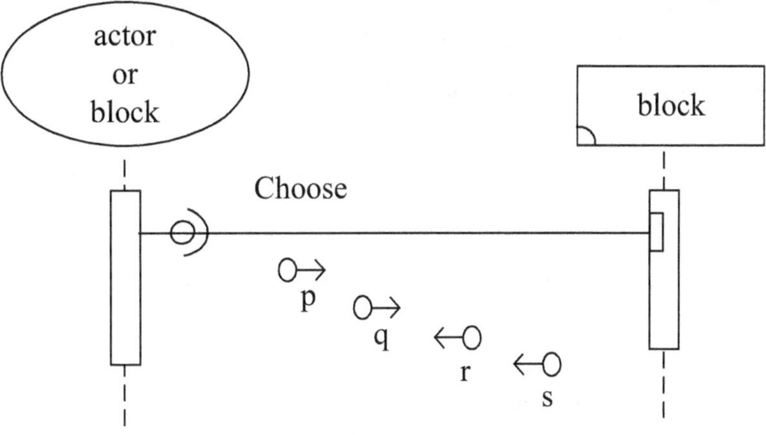

Figure 5-8. Channel-Based Value-Passed Interaction Diagram

5-4 Formal Description of Channel-Based Value-Passing Interactions

The external environment uses a "channel-based value-passing type 1 interaction" to interact with a block as shown in Figure 5-9.

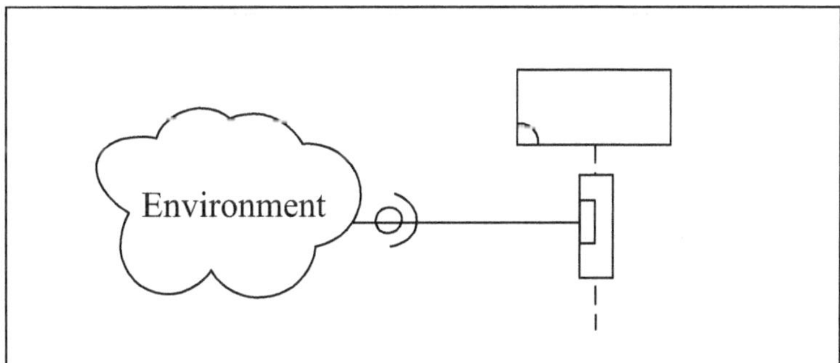

Figure 5-9. The External Environment Uses a "Type-1 Interction" to Interact with a Block

We formally describe the "channel-based value-passing type 1 interaction" as a relation $G \subseteq B \times K \times \Gamma$ where B is a set of "external environment's actors" and K is a relation of "channel signatures" and Γ is a set of "blocks".

DEFINITION (CHANNEL-BASED VALUE-PASSING TYPE 1 INTERACTION) A Channel-Based Value-Passing Type 1 Interaction CV1I = (B, K, Γ, G) consists of

- a finite set B of "external environment's actors",
- a relation K of "channel signatures",
- a finite set Γ of "blocks",
- a relation $G \subseteq B \times K \times \Gamma$, and $(\beta, k, b) \in G$.

Two blocks use a "channel-based value-passing type 2 interaction" to interact with each other as shown in Figure 5-10.

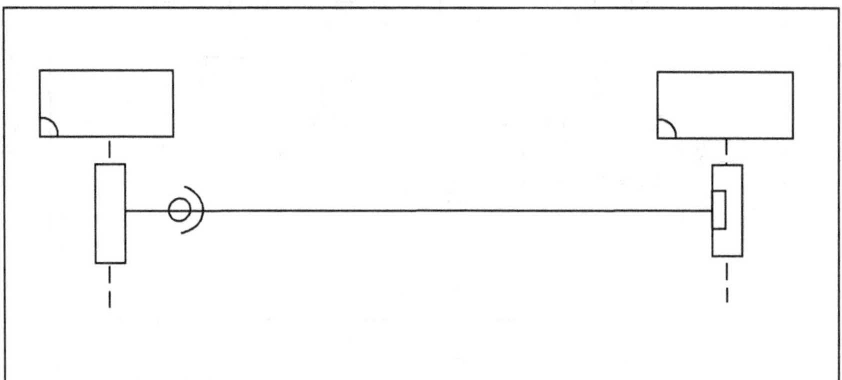

Figure 5-10. Two Blocks Use a "Type-2 Interaction" to Interact with Each Other

We formally describe the "channel-based value-passing type 2 interaction" as a relation $V \subseteq \Gamma_1 \times K \times \Gamma_2$ where Γ is a set of "blocks" and K is a relation of "channel signatures".

DEFINITION (CHANNEL-BASED VALUE-PASSING TYPE 2 INTERACTION) A Channel-Based Value-Passing Type 2 Interaction CV2I = (Γ, K, V) consists of

- a finite set Γ of "blocks",
- a relation K of "channel signatures",
- a relation $V \subseteq \Gamma_1 \times K \times \Gamma_2$, and $(b_1, k, b_2) \in V$.

The external environment or block uses a "channel-based value-passing type 1 or 2 interaction" to interact with a block as shown in Figure 5-11.

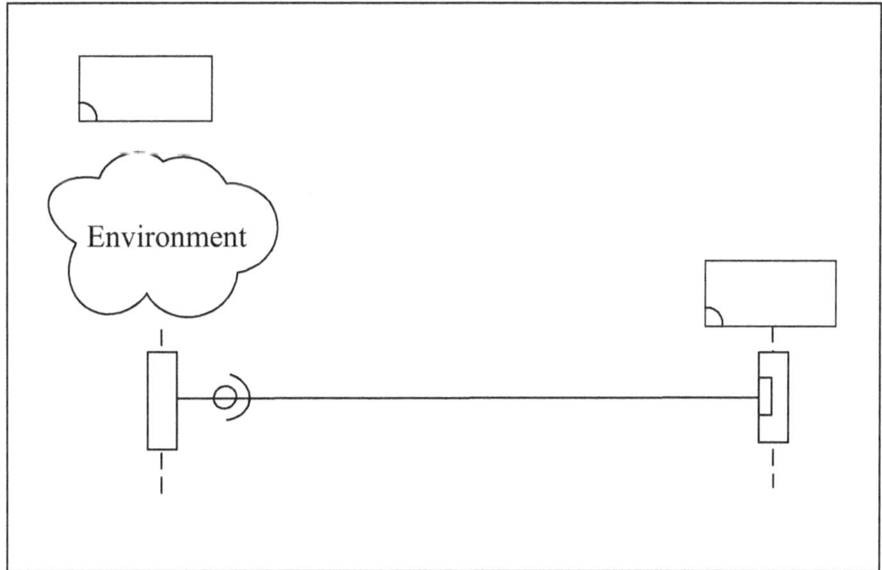

Figure 5-11. The External Environment or Block Uses a "Type-1-or-2 Interction" to Interact with a Block

We formally describe the "channel-based value-passing type 1 or 2 interaction" as a relation $\Delta \subseteq \Xi \times K \times \Gamma$ where Ξ is a set of "external environment's actors or blocks" and K is a relation of "channel signatures" and Γ is a set of "blocks".

DEFINITION (CHANNEL-BASED VALUE-PASSING INTERACTION) A Channel-Based Value-Passing Interaction CVI = (Ξ, K, Γ, Δ) consists of

- a finite set Ξ of "external environment's actors or blocks",
- a relation K of "channel signatures",
- a finite set Γ of "blocks",
- a relation $\Delta \subseteq \Xi \times K \times \Gamma$, and $(\rho, k, b) \in \Delta$.

Chapter 6: Operation-Based Value-Passing Interactions

In this chapter, we first introduce operations and operation-based value-passing interactions. We then introduce the formal description of operation-based communication ports, operation-based value-passing actions and operation-based value-passing interactions.

6-1 Operations

A block can **provide** multiple operations that represent the procedure, method, or function of the block. We use a component operation diagram (COD), as shown in Figure 6-1, to display the operations of all blocks in the system.

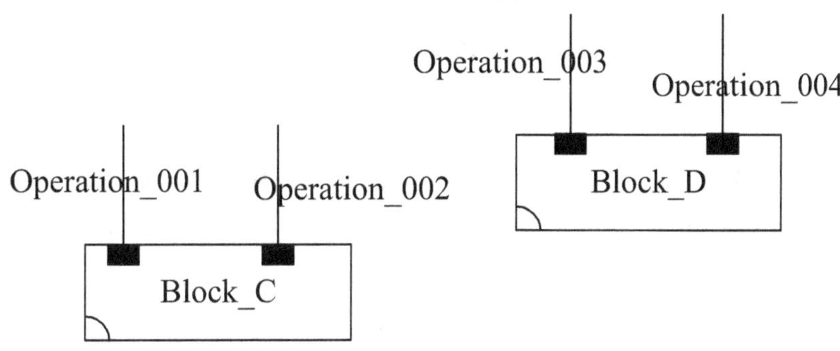

Figure 6-1. Component Operation Diagram

An operation may contain several input parameters (e.g. i_1, i_2) and output parameters (e.g. o_1, o_2), as shown in Figure 6-2.

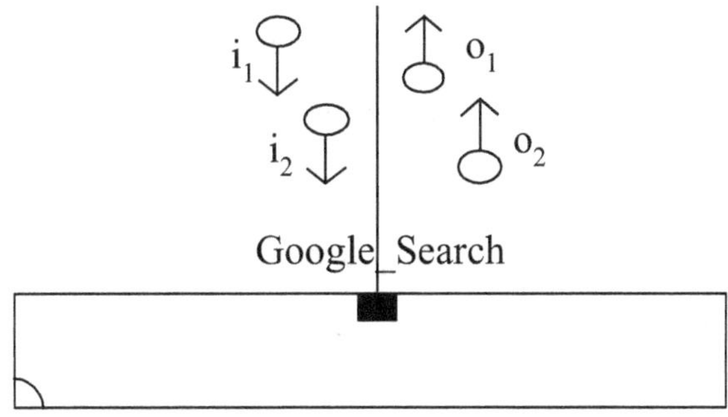

Figure 6-2. An Operation Contains Several Input/Output Parameters

An operation (can be extended to operation call or operation return) signature is used to completely describe an operation. The signature for an operation is a combination of its name along with parameters as follows:

<operation name> (<parameter list>)

The parameters in the parameter list represent the inputs or outputs of the operation. Each parameter in the list is displayed with the following format:

<direction> <parameter name> : <parameter type>

Parameter direction may be in, out, or inout. We formally describe the "operation call or operation return signature" as a relation $L \subseteq \Lambda \times \Theta$ where Λ is a set of "operation names" and Θ is a set of "parameter lists".

DEFINITION (OPERATION CALL OR OPERATION RETURN SIGNATURE) An Operation Call or Operation Return Signature OS = (Λ, Θ, L) consists of

. a finite set Λ of "operation names",
. a finite set Θ of "parameter lists",
. a relation $L \subseteq \Lambda \times \Theta$, and $(op, p) \in L$.

6-2 Component Operation Diagram

We formally describe the "component operation diagram" as a relation $COD \subseteq L \times \Gamma$ where L is a relation of "operation signatures" and Γ is a set of "blocks".

DEFINITION (COMPONENT OPERATION DIAGRAM) A Component Operation Diagram COD = (L, Γ, COD) consists of

- a relation L of " operation signatures",
- a finite set Γ of "blocks",
- a relation $COD \subseteq L \times \Gamma$, and $(l, b) \in COD$.

6-3 Operation-Based Value-Passing Interactions

An interaction represents an indivisible and instantaneous handshake or communication between two agents. In the operation-based value-passing approach as shown in Figure 6-3, the caller agent (either external environment's actor or block) communicates with the callee agent (block) through the operation call (solid line) or operation return (dashed line) interaction (also known as operation call or operation reply message)... In the figure, "Google_Search (In a, b)" is an operation call signature and "Google_Search (Out c, d)" is an operation return signature. The operation call signature and its corresponding operation return signature can be merged into an operation signature (i.e., Google_Search (In a, b; Out c, d). The figure also depicts that the "Google_Search" operation is **required** by the caller agent and is **provided** by the callee agent.

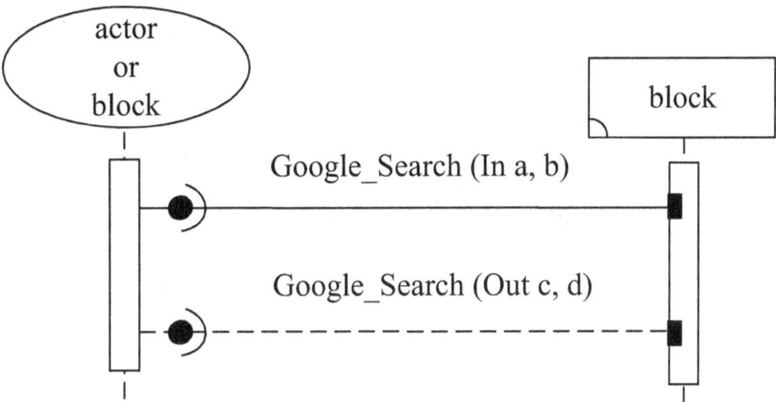

Figure 6-3. Operation-Based Value-Passing Interactions

The caller agent owns the "calling port" of the interaction. In the operation call interaction (also known as operation call message) case, the calling port is "$\overline{\text{Google\_Search (In a, b)}}$" and its conduct is to assist the caller agent to output a value (send a message) to each of the "a" and "b" variables (of the "Google_Search" operation), as shown in Figure 6-4.

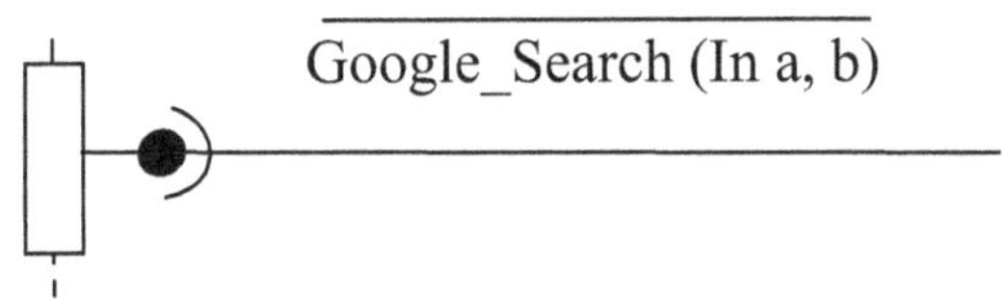

Figure 6-4. Calling Port in the Operation Call Interaction Case

The caller agent together with the "calling port" is named the "calling action" as shown in Figure 6-5.

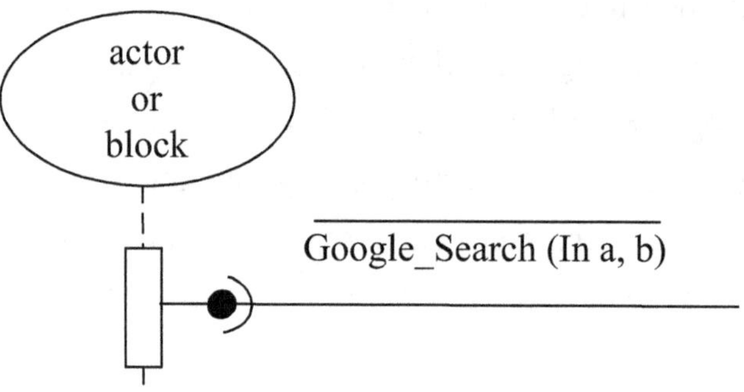

Figure 6-5. Calling Action in the Operation Call Interaction Case

In the operation return interaction (also known as operation reply message) case, the calling port is "$\overline{\text{Google\_Search (Out c, d)}}$" and its conduct is to assist the caller agent to input a value (receive a message) from each of the "c" and "d" variables (of the "Google_Search" operation), as shown in Figure 6-6.

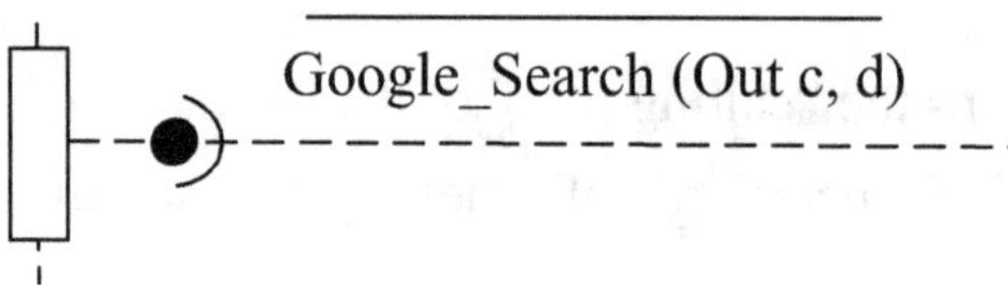

Figure 6-6. Calling Port in the Operation Return Interaction Case

The caller agent together with the "calling port" is named the "calling action" as shown in Figure 6-7.

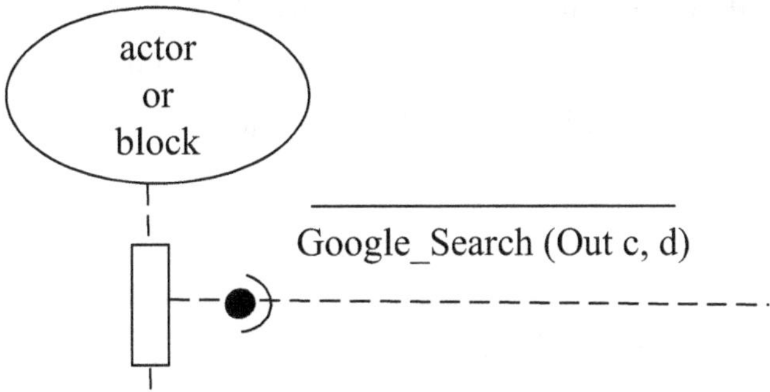

Figure 6-7. Calling Action in the Operation Return Interaction Case

The callee agent owns the "called port" of the interaction. In the operation call interaction case, the called port is "Google_Search (In a, b)" and its conduct is to assist the callee agent to input a value (receive a message) for each of the "a" and "b" variables (of the "Google_Search" operation), as shown in Figure 6-8.

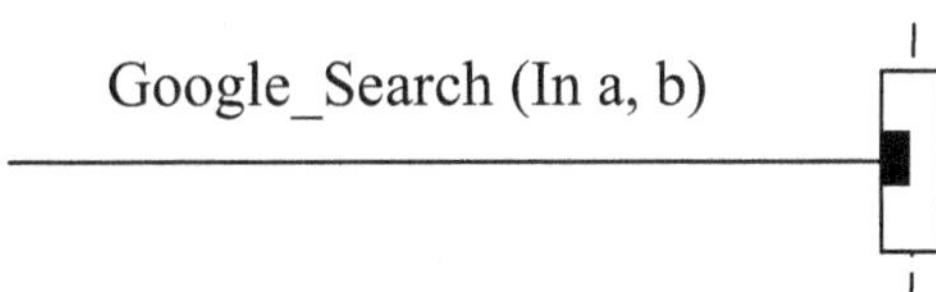

Figure 6-8. Called Port in the Operation Call Interaction Case

The callee agent together with the "called port" is named the "called action" as shown in Figure 6-9.

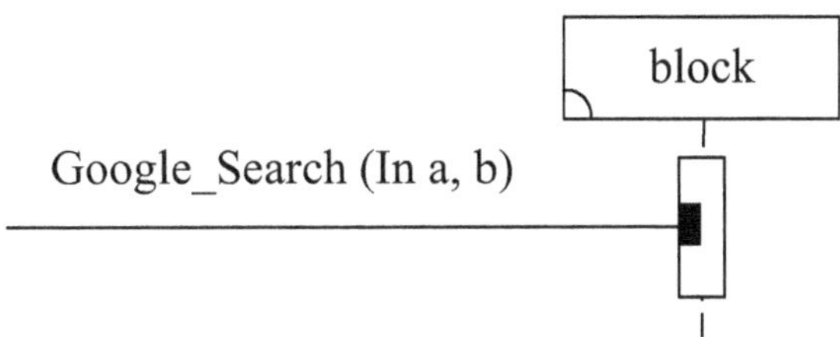

Figure 6-9. Called Action in the Operation Call Interaction Case

In the operation return interaction case, the called port is "Google_Search (Out c, d)" and its conduct is to assist the callee agent to output a value (send a message) for each of the "c" and "d" variables (of the "Google_Search" operation), as shown in Figure 6-10.

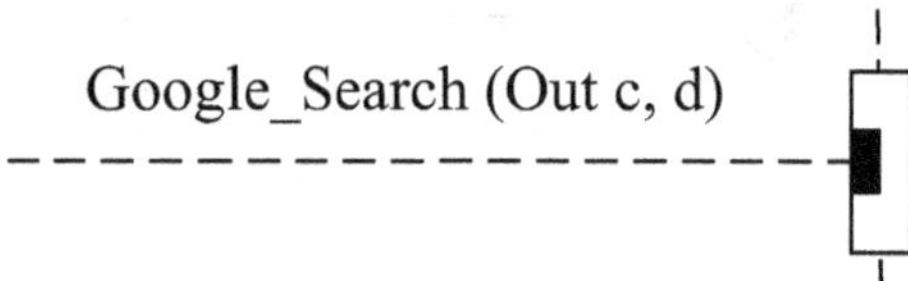

Figure 6-10. Called Port in the Operation Return Interaction Case

The callee agent together with the "called port" is named the "called action" as shown in Figure 6-11.

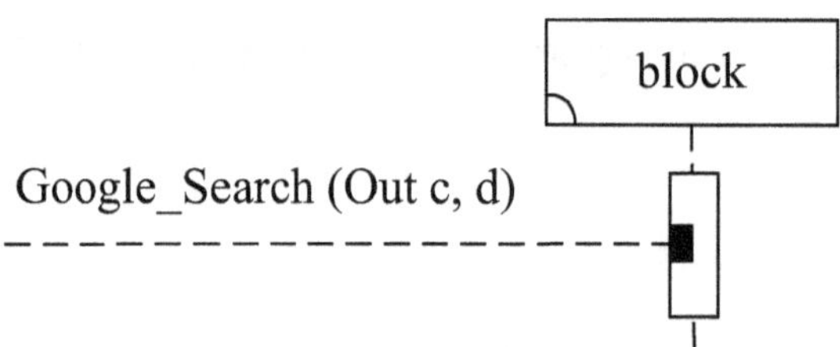

Figure 6-11. Called Action in the Operation Return Interaction Case

Also we can draw the "operation-based value-passing interaction" diagram as shown in Figure 6-12.

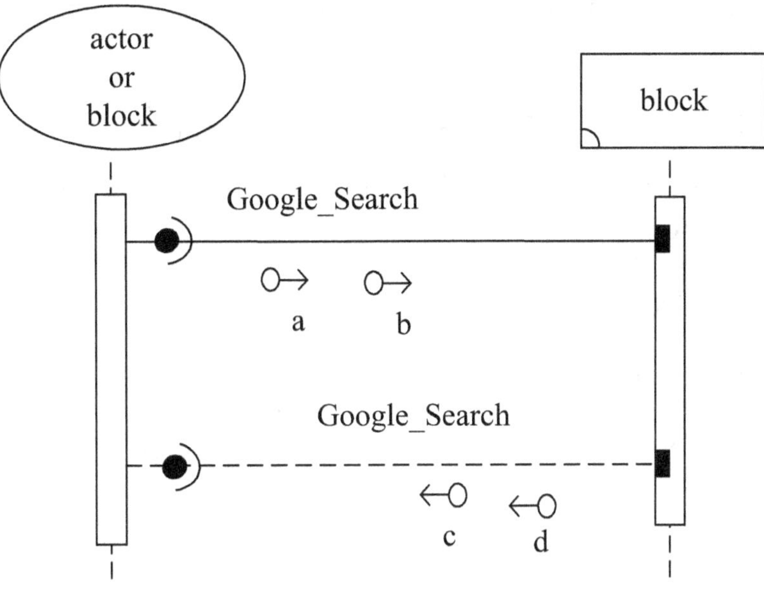

Figure 6-13. Operation-Based Value-Passing Interaction Diagram

6-4 Formal Description of Operation-Based Value-Passing Interactions

The external environment uses an "operation-based value-passing type 1 interaction" to interact with a block as shown in Figure 6-13.

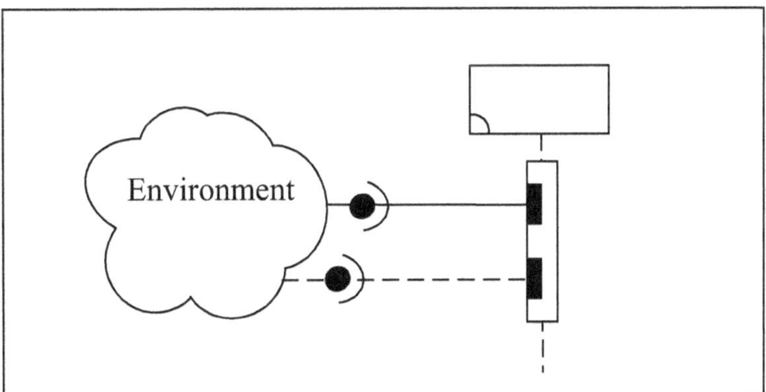

Figure 6-13. The External Environment Uses a "Type-1 Interction" to Interact with a Block

We formally describe the "operation-based value-passing type 1 interaction" as a relation $G \subseteq N \times B \times L \times \Gamma$ where N is a set of "operation call or operation return tags" and B is a set of "external environment's actors" and L is a relation of "operation call or operation return signatures" and Γ is a set of "blocks".

DEFINITION (OPERATION-BASED VALUE-PASSING TYPE 1 INTERACTION) An Operation-Based Value-Passing Type 1 Interaction OV1I = (N, B, L, Γ, G) consists of

- a finite set N of "operation call or operation return tags",
- a finite set B of "external environment's actors",
- a relation L of " operation call or operation return signatures",
- a finite set Γ of "blocks",
- a relation $G \subseteq N \times B \times L \times \Gamma$, and $(n, \beta, l, b) \in G$.

Two blocks use an "operation-based value-passing type 2 interaction" to interact with each other as shown in Figure 6-14.

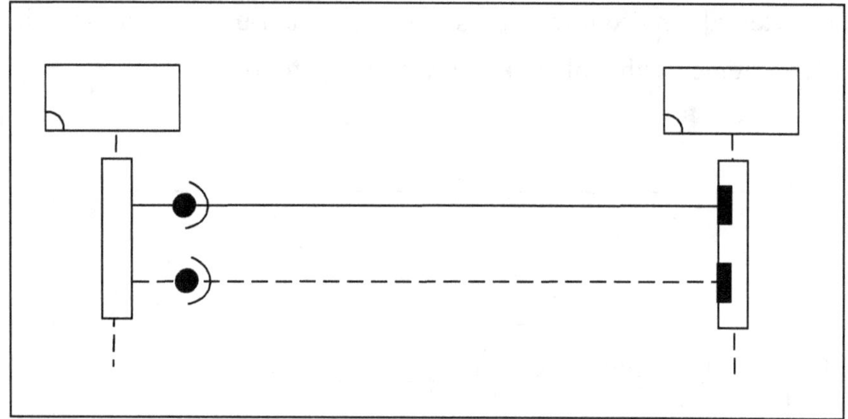

Figure 6-14. Two Blocks Use a "Type-2 Interaction" to Interact with Each Other

We formally describe the "operation-based value-passing type 2 interaction" as a relation $V \subseteq N \times \Gamma_1 \times L \times \Gamma_2$ where N is a set of "operation call or operation return tags" and Γ is a set of "blocks" and L is a relation of "operation call or operation return signatures".

DEFINITION (OPERATION-BASED VALUE-PASSING TYPE 2 INTERACTION) An Operation-Based Value-Passing Type 2 Interaction OV2I = (N, Γ, L, V) consists of

- a finite set N of "operation call or operation return tags",
- a finite set Γ of "blocks",
- a relation L of " operation call or operation return signatures",
- a relation $V \subseteq N \times \Gamma_1 \times L \times \Gamma_2$, and $(n, b_1, l, b_2) \in V$.

The external environment or block uses an "operation-based value-passing type 1 or 2 interaction" to interact with a block as shown in Figure 6-15.

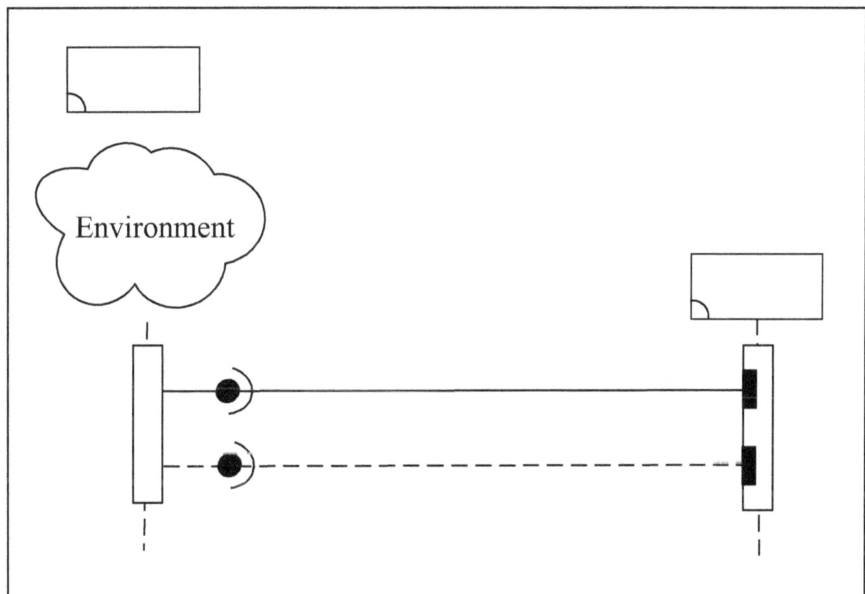

Figure 6-15. The External Environment or Block Uses a "Type-1-or-2 Interction" to Interact with a Block

We formally describe the "operation-based value-passing (type 1 or 2) interaction" as a relation $\Delta \subseteq N \times \Xi \times L \times \Gamma$ where N is a set of "operation call or operation return tags" and Ξ is a set of "external environment's actors or blocks" and L is a relation of "operation call or operation return signatures" and Γ is a set of "blocks".

DEFINITION (OPERATION-BASED VALUE-PASSING INTERACTION) An

Operation-Based Value-Passing Interaction OVI = $(N, \varXi, L, \varGamma, \varDelta)$ consists of

- a finite set N of “operation call or operation return tags”,
- a finite set $\varXi$ of “external environment’s actors or blocks”,
- a relation L of ” operation call or operation return signatures”,
- a finite set $\varGamma$ of “blocks”,
- a relation $\varDelta \subseteq N \times \varXi \times L \times \varGamma$, and $(n, \rho, l, b) \in \varDelta$.

Chapter 7: Internal Interaction or Non-Operable Interaction

In addition to channel-based value-passing interaction and operation-based value-passing interaction, there are many more interactions. The first is internal interaction, and the second is non-operable interaction.

In this chapter, we first introduce internal interactions. Then, we introduce the non-operable interaction.

7-1 Internal Interactions

If the caller agent and the callee agent are the same block, the internal interaction denoted as ”λ” will be used to represent the communication or handshake between them.

In the channel-based value-passing case, an example of internal interaction is shown in Figure 7-1.

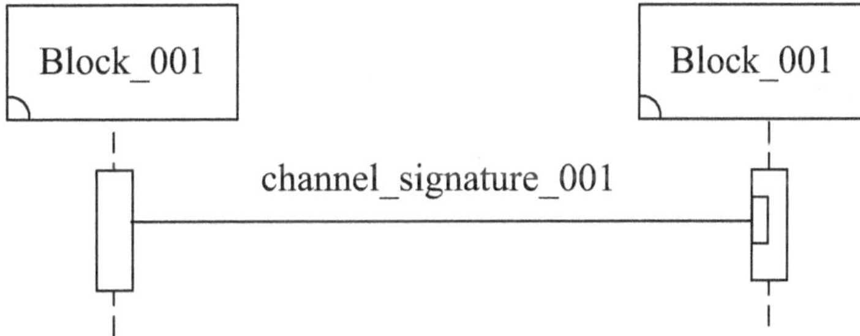

Figure 7-1. "λ" Interaction in the Channel-Based Value-Passed Case

In the operation-based value-passing case, an example of internal interaction is shown in Figure 7-2.

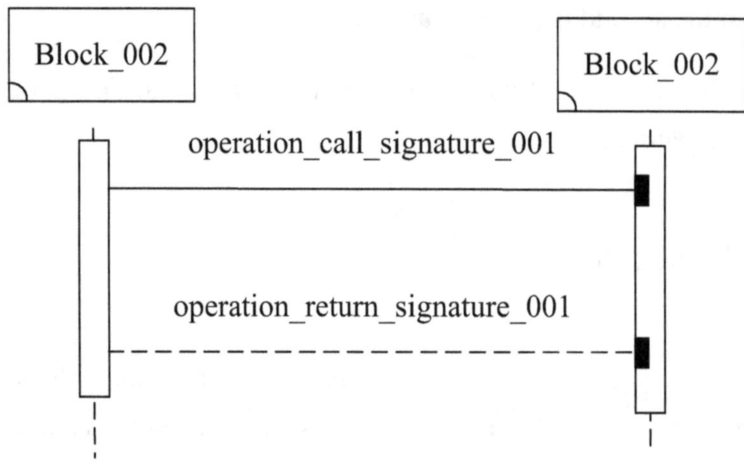

Figure 7-2. "λ" Interaction in the Operation-Based Value-Passed Case

7-2 Non-Operable Interactions

A non-operable Interaction, denoted by "*NOI*", means that it is an interaction which does nothing in the SBC state machine.

Chapter 8: Prefixes

As we recall that SBC state machine is a labelled transition system (LTS) which provides a single diagram to integrate structural and behavioral constructs in the MBSE modeling construction. In SBC state machine, each transition is labelled with a prefix.

In this chapter, we first introduce operation-based value-passing related interactions. We then introduce the guarded condition. Last, we introduce the definition of prefix.

8-1 Operation-Based Value-Passing Related Interaction

Before giving the definition of the prefix, we first need to define operation-based value-passing related interactions.

DEFINITION (OPERATION-BASED VALUE-PASSING RELATED INTERACTION) An Operation-Based Value-Passing Related Interaction $OVRI = (\lambda, NOI, \Delta, \Omega)$ consists of

- an internal interaction "λ",
- a non-operable interaction "*NOI* ",
- a relation Δ of " operation-based value-passing interactions",
- a finite set $\Omega = \{\lambda\} \cup \{NOI\} \cup \Delta$.

8-2 Guard Conditions

In SBC state machine, guard conditions (or simply guards) are Boolean expressions evaluated dynamically based on the value of variables. Guard conditions affect the behavior of a SBC state machine by enabling interactions only when they evaluate to TRUE and disabling them when they evaluate to FALSE.

8-3 Definition of Prefix

In SBC state machine, each transition is labelled with a prefix. We are now ready to give the definition of the prefix.

DEFINITION (PREFIX) A Prefix $PX = (C, \Omega, \Pi, R)$ consists of

- a finite set C of optional guard conditions,
- a finite set Ω of operation-based value-passing related interactions,
- a finite set Π of optional code snippets
- a relation $R \subseteq C \times \Omega \times \Pi$, and $(c, \alpha, \pi) \in R$.

In SBC state machine, all prefixes are guarded by either an explicitly given condition, or the implicit condition [TRUE]. If the value of the condition is TRUE and the interaction is ready, the transition will be triggered. Once the transition is triggered, the appended code snippet will be executed.

Chapter 9: Generalized SBC State Machines

SBC state machine is a prefix-labelled transition system which provides a single diagram for MBSE to integrate structural and behavioral constructs. In the SBC state machine, each state expression is regarded as a process. The SBC state machine has three different levels of complexity: basic SBC state machine, intermediate state machine and advanced state machine.

In order to integrate these three different SBC state machines, we need to introduce a generalized SBC state machine that shares the common definitions of basic SBC state machines, intermediate SBC state machines and advanced SBC state machines.

In this chapter, we first introduce the definition of generalized SBC state machine. We then introduce the graphical and relational representation and execution of SBC state machine. We will also show how to rename all states of the SBC state machine. Last, we introduce the orthogonal composite state and inactive state.

9-1 Definition of Generalized SBC State Machine

We are now ready to give the definition of generalized SBC state machine. The notion of a generalized SBC state machine is defined as follows.

DEFINITION (GENERALIZED SBC STATE MACHINE) A Generalized SBC State Machine $SSM = (\Psi, (\pi_0, s_0), R, SSMR)$ consists of

- a finite non-empty set Ψ of states,
- an optional code snippet π_0 in the initial transition, and $\pi_0 \in \Pi$,
- an initial state $s_0 \in \Psi$,
- a relation R of prefix,
- a transition relation $SSMR \subseteq \Psi_1 \times R \times \Psi_2$, where $(s_j, r, s_k) \in SSMR$ is denoted by

$$s_j \xrightarrow{r} s_k.$$

9-2 Graphical Representation of SBC state machine

We shall draw a single diagram to represent SBC state machine. Figure 9-1 shows the diagram of the SBC state machine "SSM_{001}". In the diagrammed SBC state machine, the state is represented by an oval box containing its name; the transition from the source state to the target state is indicated by an arrow labelled with a prefix; the initial state (e.g., s_{001}) is the target state of the transition that has no source state. The transition without a source state is called the initial transition. In a state, if multiple transitions to be triggered are met, the choice of trigger will be arbitrary and fair.

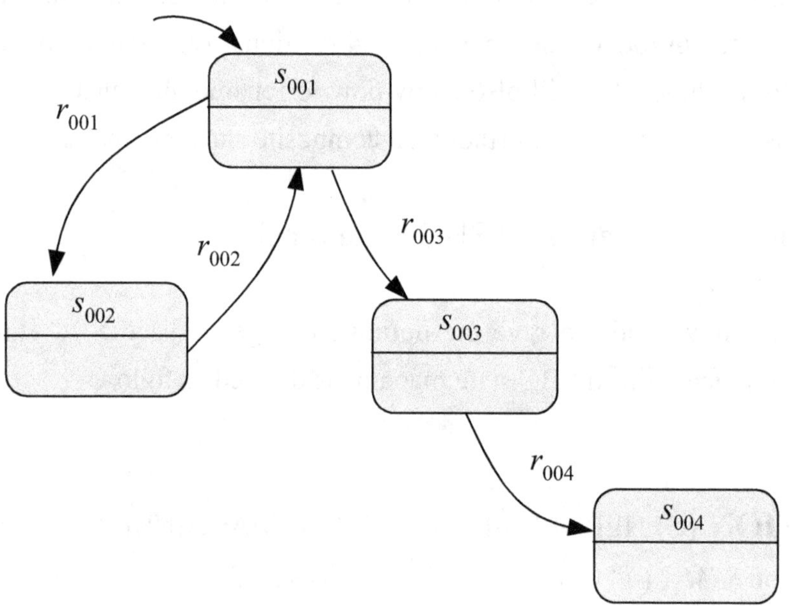

Figure 9-1. Diagram of the SBC State Machine SSM_{001}

9-3 Relational Representation of SBC State Machine

We can also list the relationships that represent SBC state machine. Figure 9-2 shows the transition relation $SSMR_{001}$ of the SBC state machine SSM_{001}.

S_1	R	S_2
s_{001}	r_{001}	s_{002}
s_{002}	r_{002}	s_{001}
s_{001}	r_{003}	s_{003}
s_{003}	r_{004}	s_{004}

Figure 9-2. Relation $SSMR_{001}$ of the SBC State Machine SSM_{001}

9-4 Renaming all States of a SBC State Machine

We say that a function f from state names to state names is a renaming function. For each renaming function f, the renaming combinator $[f]$, postfixed to a SBC state machine, has the effect of renaming the state names of a SBC state machine as dictated by f. We shall often write $s_1'/s_1,\ldots, s_n'/s_n$ for the renaming function f for which $f(s_i) = s_i'$, for i = 1,…., n.

For example, Figure 9-3 shows the diagram of the SBC state machine SSM_{991} refers to $SSM_{001}[s_{991}/s_{001}, s_{992}/s_{002}, s_{993}/s_{003}, s_{994}/s_{004}]$, written as "$SSM_{991} \stackrel{\text{ref}}{=\!=} SSM_{001}[f]$" or "$SSM_{991} \stackrel{\text{ref}}{=\!=} SSM_{001}[s_{991}/s_{001}, s_{992}/s_{002}, s_{993}/s_{003}, s_{994}/s_{004}]$".

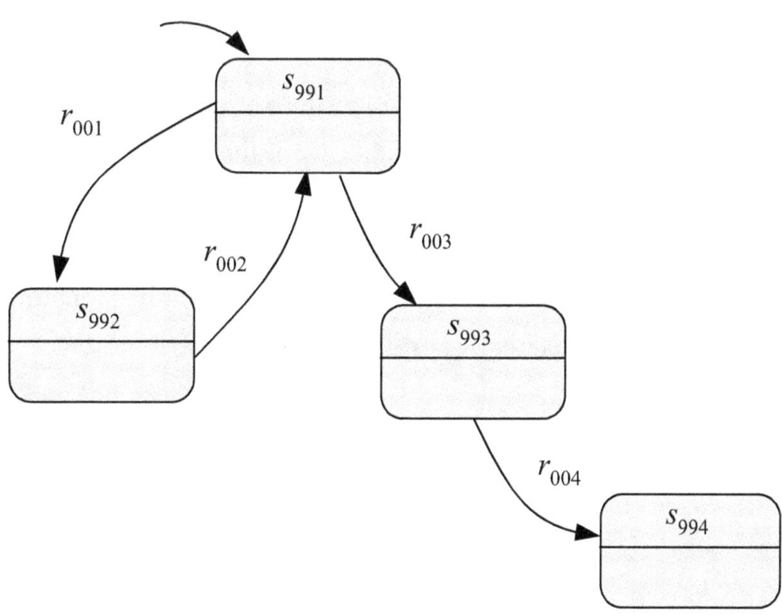

Figure 9-3. Diagram of the SBC State Machine $SSM_{001}[f]$

9-5 Execution of SBC State Machines

In the SBC state machine, guard conditions (or simply guards) are Boolean expressions evaluated dynamically based on the value of variables. Guard conditions affect the behavior of a SBC state machine by enabling interactions only when they evaluate to TRUE and disabling them when they evaluate to FALSE. In the SBC notation, guard conditions are shown in square brackets (e.g., [key_count > 0] in Figure 9-4).

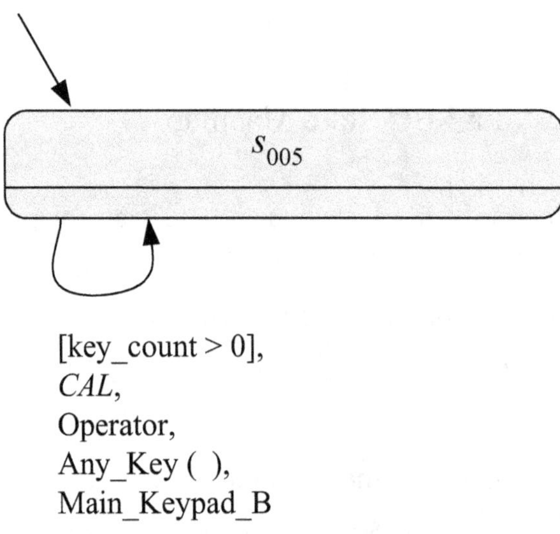

Figure 9-4. An Example of Guard Conditions

In the SBC state machine, the transition line describes the execution from one state to another. Each transition line is labelled with the prefix (i.e., guard condition, interaction) that causes the transition. For example, Figure 9-5 shows that in the “s_{006}“ state the ”Student” external actor can interact with the “University“ block through the “getPastDueBalance(out pastDueBalance)“ operation return signature when the guarded condition “Has_Registered = = YES“ becomes TRUE.

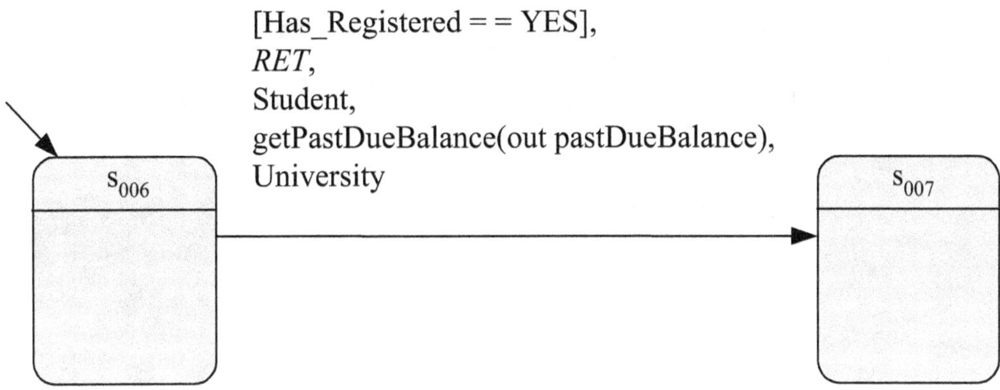

Figure 9-5. Execution of a Transition

In a state, if multiple transitions to be triggered are met, the choice of trigger will be arbitrary and fair [Quei83] (for example, Figure 9-6 shows that in the state s_{011}, the choice of prefixes r_{011}, r_{012} or r_{013} to be executed is arbitrary and fair).

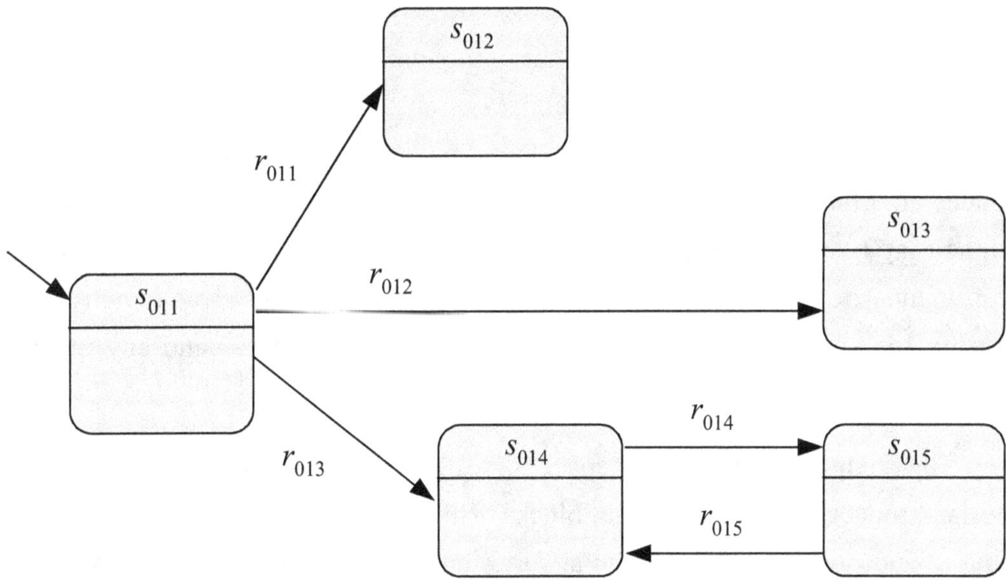

Figure 9-6. Choice of Prefixes r_{011}, r_{012} or r_{013} is Non-Deterministic

9-6 Inactive State

SBC state machine usually also include an inactive state, denoted as "*STOP*", which has no outgoing transition points as shown in Figure 9-7.

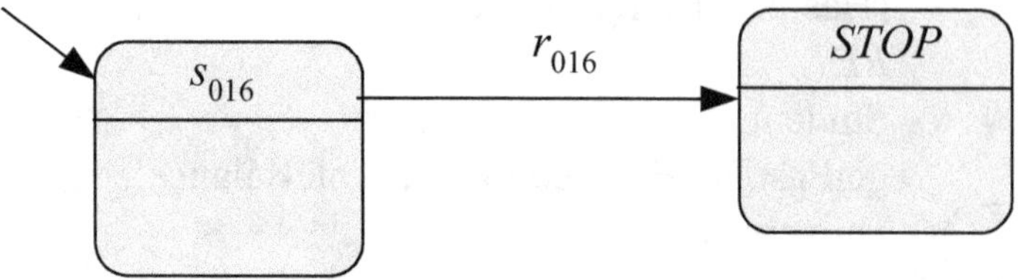

Figure 9-7. Inactive State

An inactive state is completely inactive and its unique intention is to act as the inductive anchor on top of which some interesting state expressions can be generated.

9-7 Orthogonal Composite State

In SBC state machine, states can be simple states or composite states. A simple state is one which has no substructure. A state which has substates (nested states) is called a composite state.

One special composite state is the orthogonal composite state. When a state expression involves parallel composition rules, its corresponding state space may become complicated. In order to reduce the complexity of SBC state space, we shall introduce an orthogonal composite state. An orthogonal composite state in SBC state machine may have many regions, which may each contain substates. These regions are orthogonal to each other. When an orthogonal composite state is active, each region has its own active state that is independent of the others and any incoming interaction is independently analyzed within each region.

The complete SBC state space of an orthogonal composite state is the Cartesian product of those individual blocks. The use of orthogonal regions allows the mixing of independent behaviors as a Cartesian product to be avoided and, instead, for them to remain separate. As shown in Figure 9-8, we use "$SSM_1 \| SSM_2 \| SSM_3 \| \ldots \| SSM_m$" to represent an orthogonal composite state, which means the parallel composition of SSM_1, SSM_2, SSM_3, …, and SSM_m.

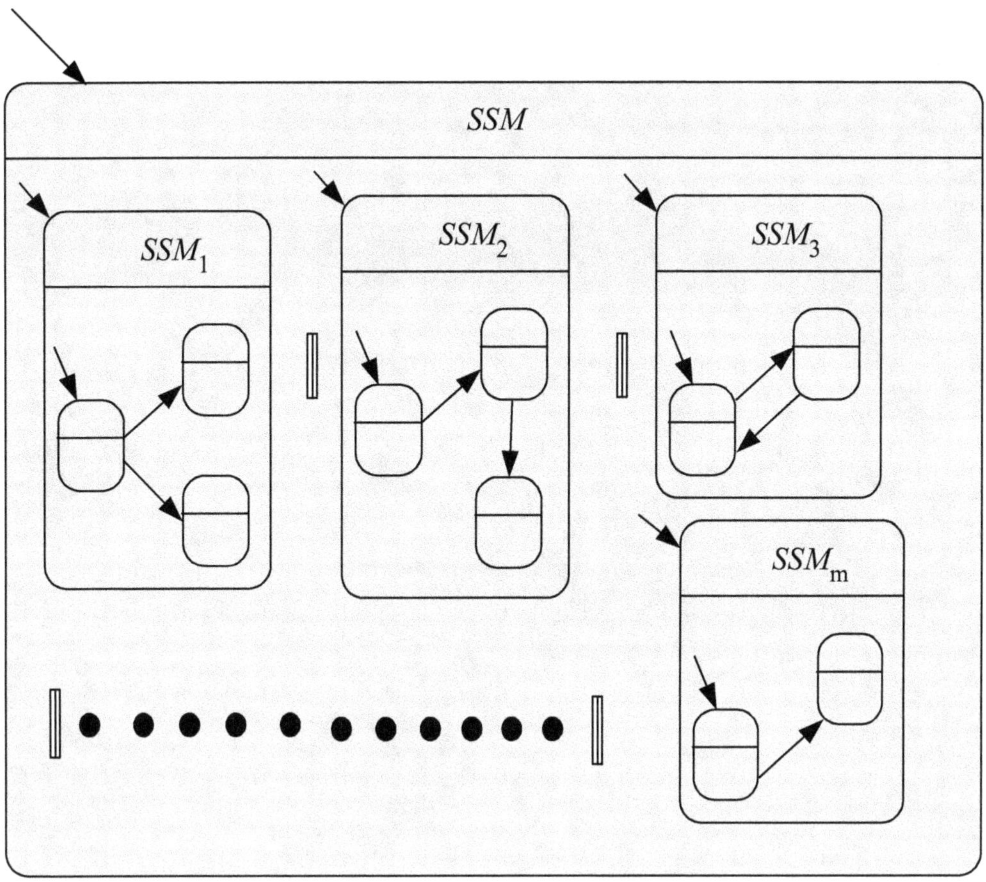

Figure 9-8. An Orthogonal Composite State of $SSM_1 \| SSM_2 \| SSM_3 \ldots \| SSM_m$

Chapter 10: Transitional Semantics of SBC State Machine

In the chapter, we detail those SBC state machine transitional semantics that play an important role in MBSE applications.

10-1 Transitional Semantics

The semantics for a SBC state machine consists in the definition of each transition relation *SSMR* (i.e. $\rightarrow$) over $\Psi_1 \times R \times \Psi_2$ which is associated with the transition rules of state operators. These transition rules will follow the construct of state expressions.

As shown in Figure 10-1, we give the complete set of transition rules; the names Sequence, Recursion, Sum, Orthogonal, Constant, and Union indicate that the rules are associated respectively with Sequence Composition, Recursion, Alternative Composition, Orthogonal Composition, Constants and Union.

Sequence

$$\frac{}{r \bullet s \xrightarrow{r} s}$$

Recursion

$$\frac{X=(z\{\mathbf{fix}(X=z)\,/X\}) \xrightarrow{r} s'}{\mathbf{fix}(X=z) \xrightarrow{r} s'}$$

Sum_j

$$\frac{s_j \xrightarrow{r} s_j'}{\sum_{i \in I} s_i \xrightarrow{r} s_j'} (j \in I)$$

Orthogonal_1

$$\frac{s_1 \xrightarrow{r} s_1'}{s_1 \| s_2 \xrightarrow{r} s_1' \| s_2}$$

Orthogonal_2

$$\frac{s_2 \xrightarrow{r} s_2'}{s_1 \| s_2 \xrightarrow{r} s_1 \| s_2'}$$

Constant_1

$$\frac{s \xrightarrow{r} s' \wedge s \neq s'}{A \xrightarrow{r} s' \wedge \text{Delete } s \xrightarrow{r} s'} \quad (A \overset{\mathbf{def}}{=} q \wedge SSM_s \overset{\mathbf{ref}}{=} SSM_q[f])$$

Constant_2

$$\frac{s' \xrightarrow{r} s \wedge s \neq s'}{s' \xrightarrow{r} A \wedge \text{Delete } s' \xrightarrow{r} s} \quad (A \overset{\mathbf{def}}{=} q \wedge SSM_s \overset{\mathbf{ref}}{=} SSM_q[f])$$

Constant_3

$$\frac{s \xrightarrow{r} s' \wedge s = s'}{A \xrightarrow{r} A \wedge \text{Delete } s \xrightarrow{r} s'} \quad (A \overset{\mathbf{def}}{=} q \wedge SSM_s \overset{\mathbf{ref}}{=} SSM_q[f])$$

Figure 10-1. Transition Rules for the SBC State Machine (I)

Union$_1$

$$\frac{s_1 \xrightarrow{r} s_1' \wedge s_1 \neq s_1'}{s_1 \cup s_2 \xrightarrow{r} s_1' \wedge \text{Delete } s_1 \xrightarrow{r} s_1'}$$

Union$_2$

$$\frac{s_2 \xrightarrow{r} s_2' \wedge s_2 \neq s_2'}{s_1 \cup s_2 \xrightarrow{r} s_2' \wedge \text{Delete } s_2 \xrightarrow{r} s_2'}$$

Union$_3$

$$\frac{s_1 \xrightarrow{r} s_1' \wedge s_1 = s_1'}{s_1 \cup s_2 \xrightarrow{r} s_1 \cup s_2 \wedge \text{Delete } s_1 \xrightarrow{r} s_1'}$$

Union$_4$

$$\frac{s_2 \xrightarrow{r} s_2' \wedge s_2 = s_2'}{s_1 \cup s_2 \xrightarrow{r} s_1 \cup s_2 \wedge \text{Delete } s_2 \xrightarrow{r} s_2'}$$

Figure 10-1. Transition Rules for the SBC State Machine (II)

Union_5

$$\frac{s_1' \xrightarrow{r} s_1 \;\wedge\; s_1 \neq s_1'}{s_1' \xrightarrow{r} s_1 \cup s_2 \;\wedge\; \text{Delete } s_1' \xrightarrow{r} s_1}$$

Union_6

$$\frac{s_2' \xrightarrow{r} s_2 \;\wedge\; s_2 \neq s_2'}{s_2' \xrightarrow{r} s_1 \cup s_2 \;\wedge\; \text{Delete } s_2' \xrightarrow{r} s_2}$$

Union_7

$$\frac{s_1' \xrightarrow{r} s_1'' \;\wedge\; \text{Equivalent}(s_1'', s_1) \;\wedge\; s_1' = s_1''}{s_1 \cup s_2 \xrightarrow{r} s_1 \cup s_2 \;\wedge\; \text{Delete } s_1' \xrightarrow{r} s_1''}$$

Union_8

$$\frac{s_2' \xrightarrow{r} s_2'' \;\wedge\; \text{Equivalent}(s_2'', s_2) \;\wedge\; s_2' = s_2''}{s_1 \cup s_2 \xrightarrow{r} s_1 \cup s_2 \;\wedge\; \text{Delete } s_2' \xrightarrow{r} s_2''}$$

Figure 10-1. Transition Rules for the SBC State Machine (III)

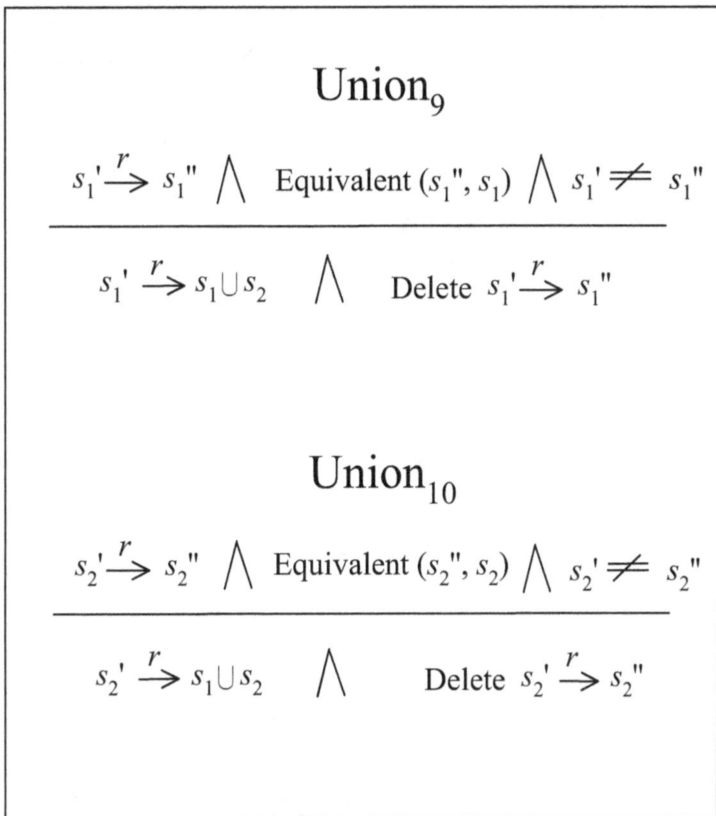

Figure 10-1. Transition Rules for the SBC State Machine (IV)

10-2 Rule of Sequence Composition

The rule for Sequence Composition, shown in Figure 10-2, can be read as follows: Under any circumstances, we always infer $r \bullet s \xrightarrow{r} s$. That is, a state expression, with a prefix prefixed to it, will use this prefix to accomplish the transition.

$$\frac{\quad}{r \bullet s \xrightarrow{r} s}$$

Figure 10-2. Rule of Sequence Composition

Let us use an example to illustrate the rule of Sequence Composition. We define the state expression "s_{021}".as "$r_{021}\bullet s_{022}$" and state expression "s_{022}".as "$r_{022}\bullet s_{023}$". Figure 10-3 shows the diagram of the SBC state machine "SSM_{021}" that represents the state expression "s_{021}".

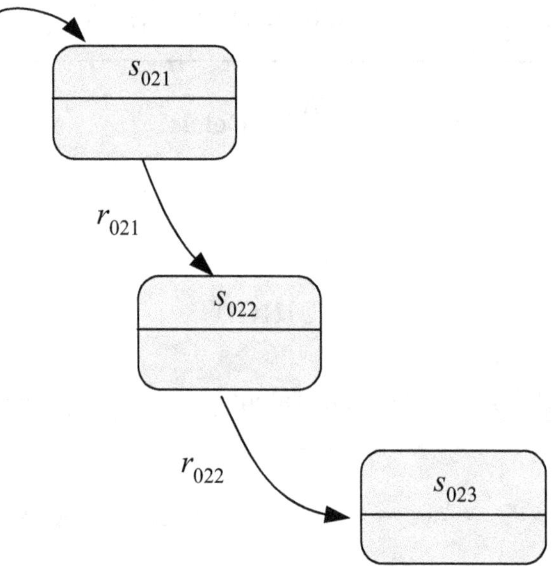

Figure 10-3. Diagram of the SBC State Machine SSM_{021}

We can also list the relationships that represent SBC state machine. Figure 10-4 shows the transition relation "$SSMR_{021}$" of the SBC state machine "SSM_{021}".

Ψ_1	R	Ψ_2
s_{021}	r_{021}	s_{022}
s_{022}	r_{022}	s_{023}

Figure 10-4. Relation $SSMR_{021}$ of the SBC State Machine SSM_{021}

10-3 Rule of Recursion

The rule for Recursion, shown in Figure 10-5, can be read as follows: This says that any prefix which may be inferred for the **fix** expression 'unwound' once (by substituting itself for its bound variable) may be inferred for the **fix** expression itself.

$$\frac{X=(z\{\mathbf{fix}(X=z)\,/X\})\xrightarrow{r} s'}{\mathbf{fix}(X=z)\xrightarrow{r} s'}$$

Figure 10-5. Rule of Recursion

Let us use an example to illustrate the rule of Recursion. We define the state expression "s_{031}".as "$\mathbf{fix}(X_{031}=r_{031}\bullet r_{032}\bullet X_{031})$". Figure 10-6 shows the diagram of the SBC state machine "SSM_{031}" which represents the state expression "s_{031}".

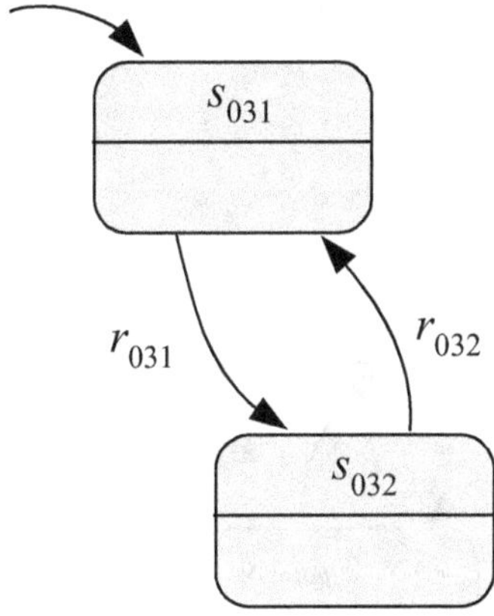

Figure 10-6. Diagram of the SBC State Machine SSM_{031}

We can also list the relationships that represent SBC state machine. Figure 10-7 shows the transition relation "$SSMR_{031}$" of the SBC state machine "SSM_{031}".

Ψ_1	R	Ψ_2
s_{031}	r_{031}	s_{032}
s_{032}	r_{032}	s_{031}

Figure 10-7. Relation $SSMR_{031}$ of the SBC State Machine SSM_{031}

For the "$r_{031}\bullet r_{032}\bullet X_{031}$" state expression, if we substitute "**fix**(X_{031}=$r_{031}\bullet r_{032}\bullet X_{031}$)" (i.e. itself) for "$X_{031}$" (i.e. its bound variable), we then get the "$r_{031}\bullet r_{032}\bullet$**fix**($X_{031}$=$r_{031}\bullet r_{032}\bullet X_{031}$)" state expression and define it as the state expression "s_{041}".

According to the rule of Recursion, the "s_{041}"state expression is equivalent to the "s_{031}"state expression. Figure 10-8 shows the diagram of the SBC state machine "SSM_{041}" which represents the state expression "s_{041}".

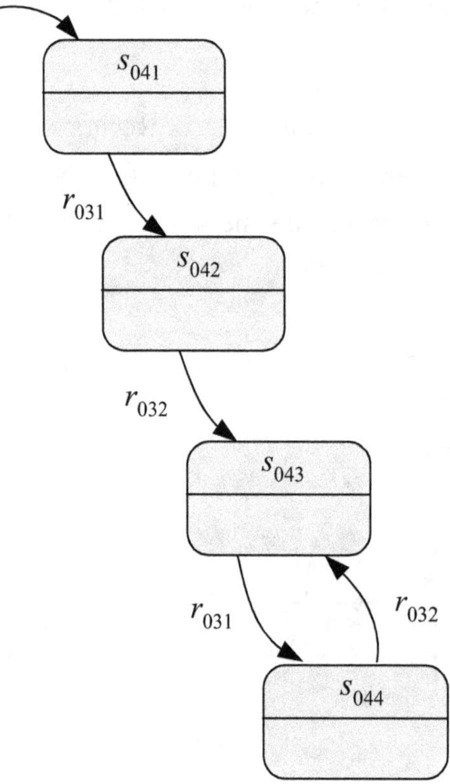

Figure 10-8. Diagram of the SBC State Machine SSM_{041}

We can also list the relationships that represent SBC state machine. Figure 10-9 shows the transition relation "$SSMR_{041}$" of the SBC state machine "SSM_{041}".

Ψ_1	R	Ψ_2
s_{041}	r_{031}	s_{042}
s_{042}	r_{032}	s_{043}
s_{043}	r_{031}	s_{044}
s_{044}	r_{032}	s_{043}

Figure 10-9. Relation $SSMR_{041}$ of the SBC State Machine SSM_{041}

10-4 Rule of Alternative Composition

The rule for Alternative Composition, shown in Figure 10-10, can be read as follows: If any one summand s_j of the sum $\sum_{i \in I} s_i$ has a prefix, then the whole sum also has that prefix.

$$\frac{s_j \xrightarrow{r} s_j'}{\sum_{i \in I} s_i \xrightarrow{r} s_j'} (j \in I)$$

Figure 10-10. Rule of Alternative Composition

Finite Alternative Composition, which is enough for many practical purposes, can be presented in a more convenient form. If I = {1, 2} then we obtain two rules for $s_1 + s_2$, by setting j = 1, 2:

$$\frac{s_1 \xrightarrow{r} s_1'}{s_1 + s_2 \xrightarrow{r} s_1'} \qquad \frac{s_2 \xrightarrow{r} s_2'}{s_1 + s_2 \xrightarrow{r} s_2'}$$

Let us use two examples to illustrate the rule of Alternative Composition. In the first example, we define the state expression "s_{051}".as "$r_{051} \bullet r_{052} \bullet s_{053}$" and state expression "$s_{061}$".as "$r_{061} \bullet s_{062}$". Figure 10-6 shows the diagram of the SBC state machine "SSM_{051}" which represents the state expression "s_{051}".

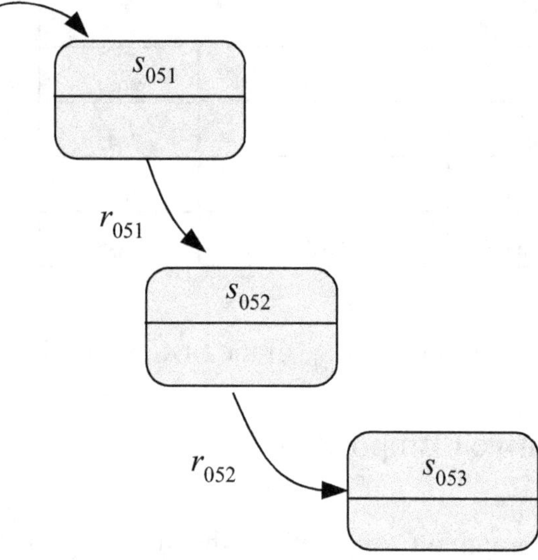

Figure 10-11. Diagram of the SBC State Machine SSM_{051}

We can also list the relationships that represent SBC state machine. Figure 10-12 shows the transition relation "$SSMR_{051}$" of the SBC state machine "SSM_{051}".

Ψ_1	R	Ψ_2
s_{051}	r_{051}	s_{052}
s_{052}	r_{052}	s_{053}

Figure 10-12. Relation $SSMR_{051}$ of the SBC State Machine SSM_{051}

Figure 10-13 shows the diagram of the SBC state machine "SSM_{061}" which represents the state expression "s_{061}".

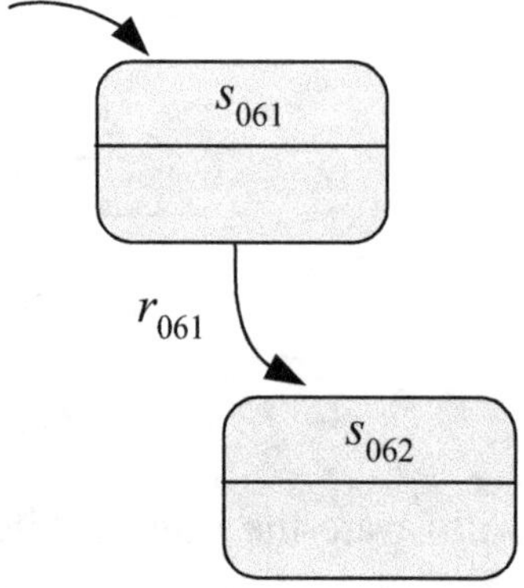

Figure 10-13. Diagram of the SBC State Machine SSM_{061}

We can also list the relationships that represent SBC state machine. Figure 10-14 shows the transition relation "$SSMR_{061}$" of the SBC state machine "SSM_{061}".

Ψ_1	R	Ψ_2
s_{061}	r_{061}	s_{062}

Figure 10-14. Relation $SSMR_{061}$ of the SBC State Machine SSM_{061}

We use the SBC state machine "SSM_{071}" to represent the state expression "$s_{051}+s_{061}$". According to the rule of Alternative Composition, the diagram of the SBC state machine "SSM_{071}" which represents the state expression "$s_{051}+s_{061}$" is shown in Figure 10-15. In the state "$s_{051}+s_{061}$", the choice of prefixes "r_{051}" and "r_{061}" to be executed is arbitrary and fair.

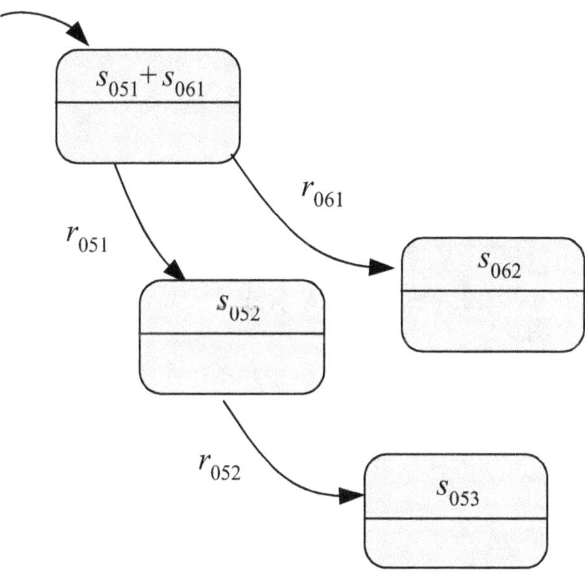

Figure 10-15. Diagram of the SBC State Machine SSM_{071}

We can also list the relationships that represent SBC state machine. Figure 10-16 shows the transition relation "$SSMR_{071}$" of the SBC state machine "SSM_{071}".

Ψ_1	R	Ψ_2
$s_{051}+s_{061}$	r_{051}	s_{052}
s_{052}	r_{052}	s_{053}
$s_{051}+s_{061}$	r_{061}	s_{062}

Figure 10-16. Relation $SSMR_{071}$ of the SBC State Machine SSM_{071}

In the second example, we define the state expression "s_{081}" as "**fix**($X_{081}=r_{081}\bullet r_{082}\bullet r_{083}\bullet X_{081}$)" and state expression "s_{091}".as "**fix**($X_{091}=r_{091}\bullet r_{092}\bullet X_{091}$)". Figure 10-17 shows the diagram of the SBC state machine "SSM_{081}" which represents the state expression "s_{081}".

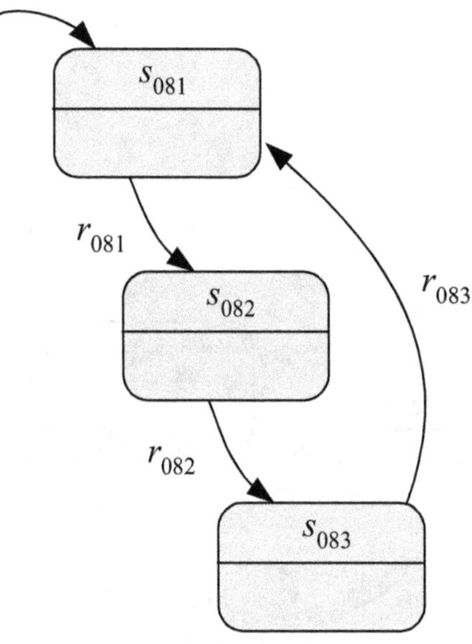

Figure 10-17. Diagram of the SBC State Machine SSM_{081}

We can also list the relationships that represent SBC state machine. Figure 10-18 shows the transition relation "$SSMR_{081}$" of the SBC state machine "SSM_{081}".

Ψ_1	R	Ψ_2
s_{081}	r_{081}	s_{082}
s_{082}	r_{082}	s_{083}
s_{083}	r_{083}	s_{081}

Figure 10-18. Relation $SSMR_{081}$ of the SBC State Machine SSM_{081}

Figure 10-19 shows the diagram of the SBC state machine "SSM_{091}" which represents the state expression "s_{091}".

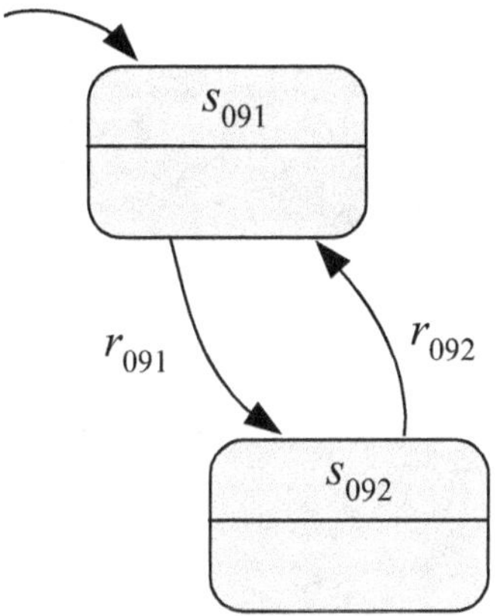

Figure 10-19. Diagram of the SBC State Machine SSM_{091}

We can also list the relationships that represent SBC state machine. Figure 10-20 shows the transition relation "$SSMR_{091}$" of the SBC state machine "SSM_{091}".

Ψ_1	R	Ψ_2
s_{091}	r_{091}	s_{092}
s_{092}	r_{092}	s_{091}

Figure 10-20. Relation $SSMR_{091}$ of the SBC State Machine SSM_{091}

We use the SBC state machine "SSM_{101}" to represent the state expression "$s_{081}+s_{091}$". According to the rule of Alternative Composition, the diagram of the SBC state machine "SSM_{101}" which represents the state expression "$s_{081}+s_{091}$" is shown in Figure 10-21. In the state "$s_{081}+s_{091}$", the choice of prefixes "r_{081}" and "r_{091}" to be executed is arbitrary and fair.

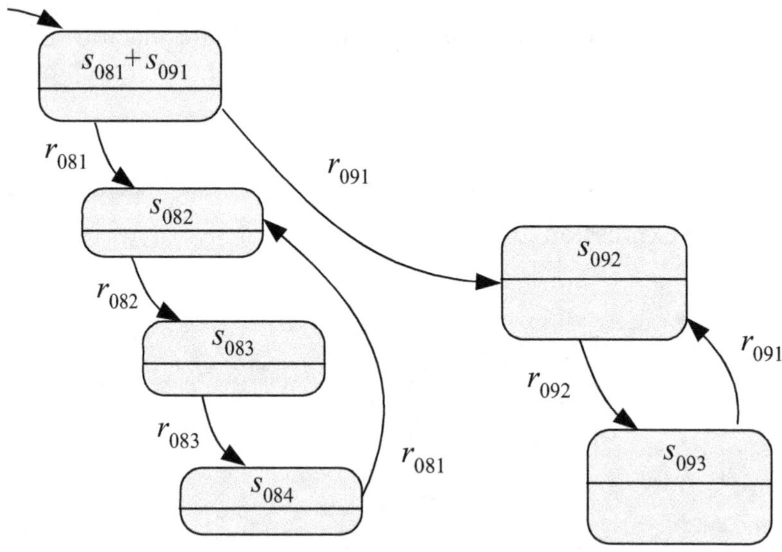

Figure 10-21. Diagram of the SBC State Machine SSM_{101}

We can also list the relationships that represent SBC state machine. Figure 10-22 shows the transition relation "$SSMR_{101}$" of the SBC state machine "SSM_{101}".

Ψ_1	R	Ψ_2
$s_{081}+s_{091}$	r_{081}	s_{082}
s_{082}	r_{082}	s_{083}
s_{083}	r_{083}	s_{084}
s_{084}	r_{081}	s_{082}
$s_{081}+s_{091}$	r_{091}	s_{092}
s_{092}	r_{092}	s_{093}
s_{093}	r_{091}	s_{092}

Figure 10-22. Relation $SSMR_{101}$ of the SBC State Machine SSM_{101}

10-5 Rule of Orthogonal Composition

There are two transition rules for Orthogonal Composition. Rule Orthogonal_1, as shown in Figure 10-23, indicates that from $s_1 \xrightarrow{r} s_1'$ we shall infer $s_1 \,\|\, s_2 \xrightarrow{r} s_1' \,\|\, s_2$.

$$\frac{s_1 \xrightarrow{r} s_1'}{s_1 \,\|\, s_2 \xrightarrow{r} s_1' \,\|\, s_2}$$

Figure 10-23. Rule Orthogonal_1

Rule Orthogonal_2, as shown in Figure 10-24, indicates that from $s_2 \xrightarrow{r} s_2'$ we shall infer $s_1 \,\|\, s_2 \xrightarrow{r} s_1 \,\|\, s_2'$.

$$\frac{s_2 \xrightarrow{r} s_2'}{s_1 \,\|\, s_2 \xrightarrow{r} s_1 \,\|\, s_2'}$$

Figure 10-24. Rule Orthogonal_2

Let us use two examples to illustrate the rule of Orthogonal Composition. In the first example, we define the state expression "s_{111}".as "$r_{111} \bullet r_{112} \bullet s_{113}$" and state expression "$s_{121}$".as "$r_{121} \bullet s_{122}$". Figure 10-25 shows the diagram of the SBC state machine "SSM_{111}" which represents the state expression "s_{111}".

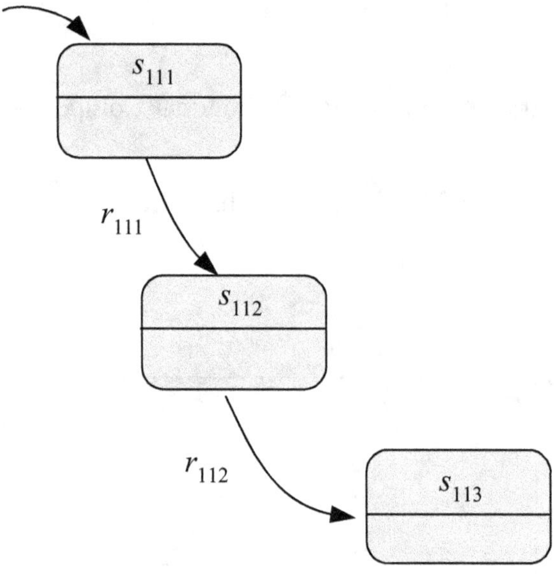

Figure 10-26. Diagram of the SBC State Machine SSM_{111}

We can also list the relationships that represent SBC state machine. Figure 10-27 shows the transition relation "$SSMR_{111}$" of the SBC state machine "SSM_{111}".

Ψ_1	R	Ψ_2
s_{111}	r_{111}	s_{112}
s_{112}	r_{112}	s_{113}

Figure 10-27. Relation $SSMR_{111}$ of the SBC State Machine SSM_{111}

Figure 10-28 shows the diagram of the SBC state machine "SSM_{121}" which represents the state expression "s_{121}".

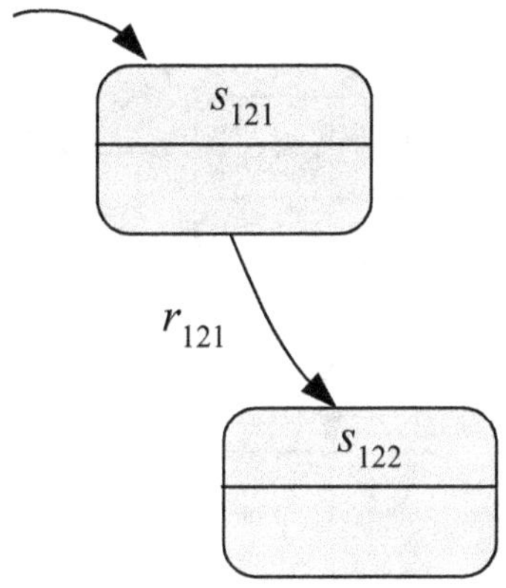

Figure 10-28. Diagram of the SBC State Machine SSM_{121}

We can also list the relationships that represent SBC state machine. Figure 10-29 shows the transition relation "$SSMR_{121}$" of the SBC state machine "SSM_{121}".

Ψ_1	R	Ψ_2
s_{121}	r_{121}	s_{122}

Figure 10-29. Relation $SSMR_{121}$ of the SBC State Machine SSM_{121}

We use the SBC state machine "SSM_{131}" to represent the state expression "$s_{111} ||| s_{121}$". According to the rule of Orthogonal Composition, the diagram of the SBC state machine "SSM_{131}" which represents the state expression "$s_{111} ||| s_{121}$" is shown in Figure 10-30.

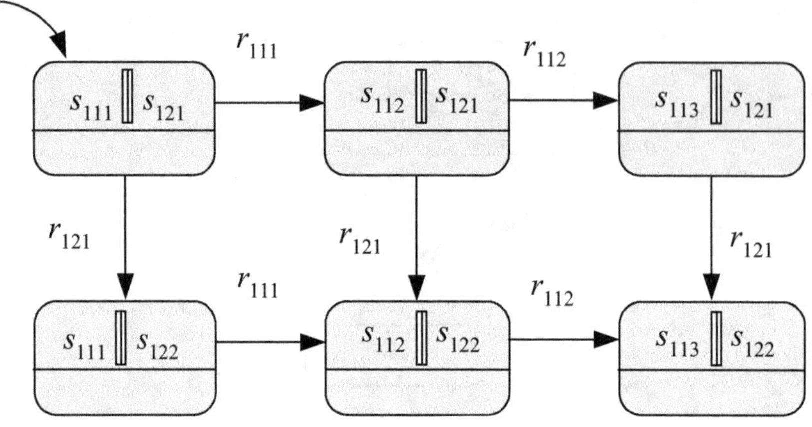

Figure 10-30. Diagram of the SBC State Machine SSM_{131}

We can also list the relationships that represent SBC state machine. Figure 10-31 shows the transition relation "$SSMR_{131}$" of the SBC state machine "SSM_{131}".

Ψ_1	R	Ψ_2
$s_{111} \parallel s_{121}$	r_{111}	$s_{112} \parallel s_{121}$
$s_{112} \parallel s_{121}$	r_{112}	$s_{113} \parallel s_{121}$
$s_{111} \parallel s_{121}$	r_{121}	$s_{111} \parallel s_{122}$
$s_{112} \parallel s_{121}$	r_{121}	$s_{112} \parallel s_{122}$
$s_{113} \parallel s_{121}$	r_{121}	$s_{113} \parallel s_{122}$
$s_{111} \parallel s_{122}$	r_{111}	$s_{112} \parallel s_{122}$
$s_{112} \parallel s_{122}$	r_{112}	$s_{113} \parallel s_{122}$

Figure 10-31. Relation $SSMR_{131}$ of the SBC State Machine SSM_{131}

In the second example, we define the state expression "s_{141}" as "**fix**($X_{141}=r_{141}\bullet r_{142}\bullet r_{143}\bullet X_{141}$)" and state expression "s_{151}".as "**fix**($X_{151}=r_{151}\bullet r_{152}\bullet X_{151}$)". Figure 10-32 shows the diagram of the SBC state machine "SSM_{141}" which represents the state expression "s_{141}".

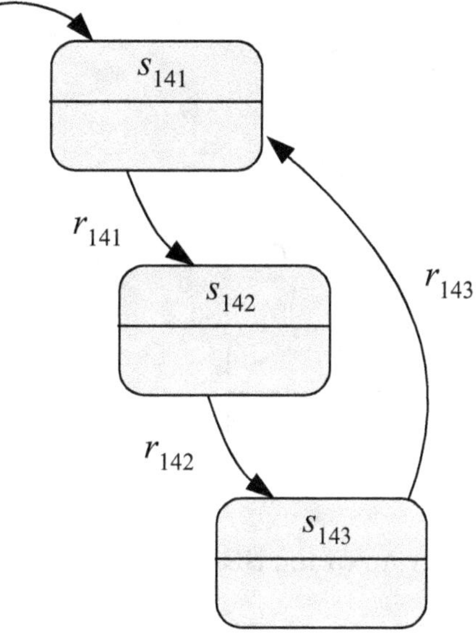

Figure 10-32. Diagram of the SBC State Machine SSM_{141}

We can also list the relationships that represent SBC state machine. Figure 10-33 shows the transition relation "$SSMR_{141}$" of the SBC state machine "SSM_{141}".

Ψ_1	R	Ψ_2
s_{141}	r_{141}	s_{142}
s_{142}	r_{142}	s_{143}
s_{143}	r_{143}	s_{141}

Figure 10-33. Relation $SSMR_{141}$ of the SBC State Machine SSM_{141}

Figure 10-34 shows the diagram of the SBC state machine "SSM_{151}" which represents the state expression "s_{151}".

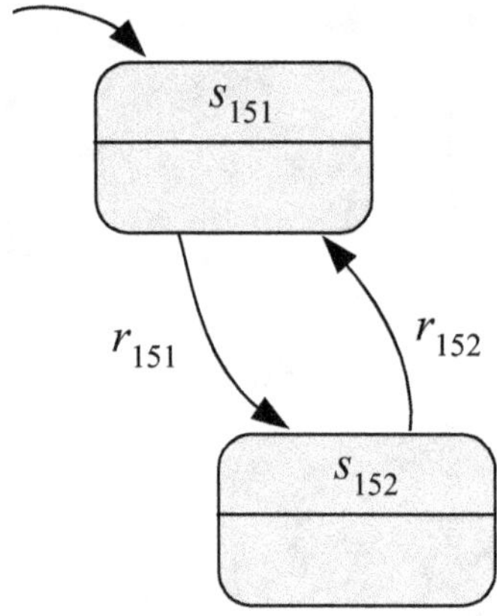

Figure 10-34. Diagram of the SBC State Machine SSM_{151}

We can also list the relationships that represent SBC state machine. Figure 10-35 shows the transition relation "$SSMR_{151}$" of the SBC state machine "SSM_{151}".

Ψ_1	R	Ψ_2
s_{151}	r_{151}	s_{152}
s_{152}	r_{152}	s_{151}

Figure 10-35. Relation $SSMR_{151}$ of the SBC State Machine SSM_{151}

We use the SBC state machine "SSM_{161}" to represent the state expression "$s_{141} \| s_{151}$". According to the rule of Alternative Composition, the diagram of the SBC state machine "SSM_{161}" which represents the state expression "$s_{141} \| s_{151}$" is shown in Figure 10-36.

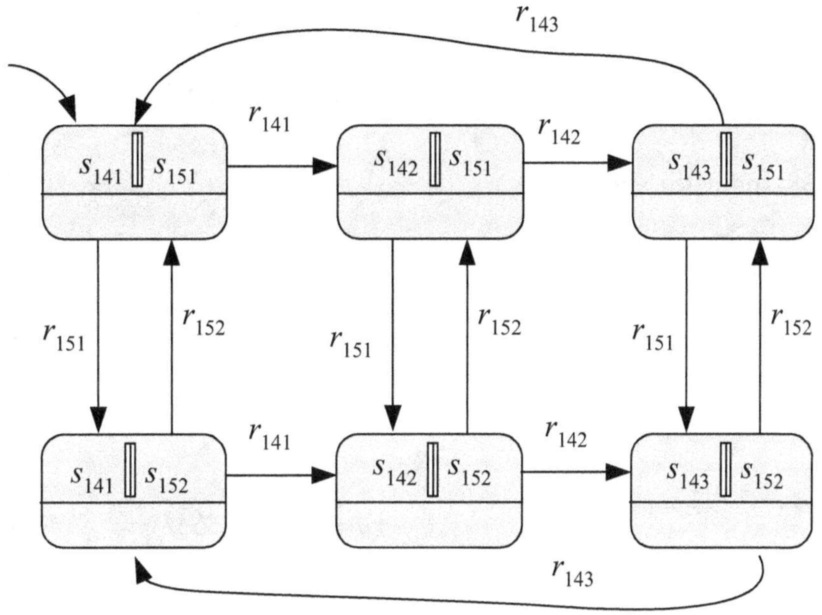

Figure 10-36. Diagram of the SBC State Machine SSM_{161}

We can also list the relationships that represent SBC state machine. Figure 10-37 shows the transition relation "$SSMR_{161}$" of the SBC state machine "SSM_{161}".

Ψ_1	R	Ψ_2
$s_{141} \parallel s_{151}$	r_{141}	$s_{142} \parallel s_{151}$
$s_{142} \parallel s_{151}$	r_{142}	$s_{143} \parallel s_{151}$
$s_{143} \parallel s_{151}$	r_{143}	$s_{141} \parallel s_{151}$
$s_{141} \parallel s_{151}$	r_{151}	$s_{141} \parallel s_{152}$
$s_{141} \parallel s_{152}$	r_{152}	$s_{141} \parallel s_{151}$
$s_{142} \parallel s_{151}$	r_{151}	$s_{142} \parallel s_{152}$
$s_{142} \parallel s_{152}$	r_{152}	$s_{142} \parallel s_{151}$
$s_{143} \parallel s_{151}$	r_{151}	$s_{143} \parallel s_{152}$
$s_{143} \parallel s_{152}$	r_{152}	$s_{143} \parallel s_{151}$
$s_{141} \parallel s_{152}$	r_{141}	$s_{142} \parallel s_{152}$
$s_{142} \parallel s_{152}$	r_{142}	$s_{143} \parallel s_{152}$
$s_{143} \parallel s_{152}$	r_{143}	$s_{141} \parallel s_{152}$

Figure 10-37. Relation $SSMR_{161}$ of the SBC State Machine SSM_{161}

10-6 Rule of Constants

There are three transition rules for Constants. Rule Constant_1, as shown in Figure 10-38, indicates that from $s \xrightarrow{r} s' \bigwedge s \neq s'$ we (1) infer $A \xrightarrow{r} s'$ and (2) delete the $s \xrightarrow{r} s'$ transition.

$$\text{Constant}_1 \qquad \frac{s \xrightarrow{r} s' \bigwedge s \neq s'}{A \xrightarrow{r} s' \bigwedge \text{Delete } s \xrightarrow{r} s'} \qquad (A \overset{\textbf{def}}{=\!=} q \bigwedge SSM_s \overset{\textbf{ref}}{=\!=} SSM_q[f])$$

Figure 10-38. Rule of Constant_1

Rule Constant_2, as shown in Figure 10-39, indicates that from $s' \xrightarrow{r} s \bigwedge s \neq s'$ we (1) infer $s' \xrightarrow{r} A$ and (2) delete the $s' \xrightarrow{r} s$ transition.

$$\text{Constant}_2 \qquad \frac{s' \xrightarrow{r} s \bigwedge s \neq s'}{s' \xrightarrow{r} A \bigwedge \text{Delete } s' \xrightarrow{r} s} \qquad (A \overset{\textbf{def}}{=\!=} q \bigwedge SSM_s \overset{\textbf{ref}}{=\!=} SSM_q[f])$$

Figure 10-39. Rule of Constant_2

Rule Constant_3, as shown in Figure 10-40, indicates that from $s \xrightarrow{r} s' \bigwedge s = s'$ we (1) infer $A \xrightarrow{r} A$ and (2) delete the $s \xrightarrow{r} s'$ transition.

Constant$_3$

$$\frac{s \xrightarrow{r} s' \wedge s = s'}{A \xrightarrow{r} A \wedge \text{Delete } s \xrightarrow{r} s'} \quad (A \overset{\textbf{def}}{=\!=} q \wedge SSM_s \overset{\textbf{ref}}{=\!=} SSM_q[f])$$

Figure 10-40. Rule of Constant$_3$

10-7 Rule of Union

There are ten transition rules for Union. Rule Union$_1$, as shown in Figure 10-41, indicates that from $s_1 \xrightarrow{r} s_1' \wedge s_1 \neq s_1'$ we (1) infer $s_1 \cup s_2 \xrightarrow{r} s_1'$ and (2) delete the $s_1 \xrightarrow{r} s_1'$ transition.

$$\frac{s_1 \xrightarrow{r} s_1' \wedge s_1 \neq s_1'}{s_1 \cup s_2 \xrightarrow{r} s_1' \quad \wedge \quad \text{Delete } s_1 \xrightarrow{r} s_1'}$$

Figure 10-41. Rule Union$_1$

Rule Union$_2$, as shown in Figure 10-42, indicates that from $s_2 \rightarrow s_2' \wedge s_2 \neq s_2'$ we (1) infer $s_1 \cup s_2 \xrightarrow{r} s_2'$ and (2) delete the $s_2 \xrightarrow{r} s_2'$ transition.

$$\frac{s_2 \xrightarrow{r} s_2' \ \wedge \ s_2 \neq s_2'}{s_1 \cup s_2 \xrightarrow{r} s_2' \ \wedge \ \text{Delete } s_2 \xrightarrow{r} s_2'}$$

Figure 10-42. Rule Union_2

Rule Union_3, as shown in Figure 10-43, indicates that from $s_1 \xrightarrow{r} s_1' \wedge s_1 = s_1'$ we (1) infer $s_1 \cup s_2 \xrightarrow{r} s_1 \cup s_2$ and (2) delete the $s_1 \xrightarrow{r} s_1'$ transition.

$$\frac{s_1 \xrightarrow{r} s_1' \ \wedge \ s_1 = s_1'}{s_1 \cup s_2 \xrightarrow{r} s_1 \cup s_2 \ \wedge \ \text{Delete } s_1 \xrightarrow{r} s_1'}$$

Figure 10-43. Rule Union_3

Rule Union_4, as shown in Figure 10-44, indicates that from $s_2 \xrightarrow{r} s_2' \wedge s_2 = s_2'$ we (1) infer $s_1 \cup s_2 \xrightarrow{r} s_1 \cup s_2$ and (2) delete the $s_2 \xrightarrow{r} s_2'$ transition.

$$\frac{s_2 \xrightarrow{r} s_2' \ \wedge \ s_2 = s_2'}{s_1 \cup s_2 \xrightarrow{r} s_1 \cup s_2 \ \wedge \ \text{Delete } s_2 \xrightarrow{r} s_2'}$$

Figure 10-44. Rule Union_4

Rule Union5, as shown in Figure 10-45, indicates that from $s_1' \xrightarrow{r} s_1 \wedge s_1 \neq s_1'$, we (1) infer $s_1' \xrightarrow{r} s_1 \cup s_2$ and (2) delete the $s_1' \xrightarrow{r} s_1$ transition.

$$\frac{s_1' \xrightarrow{r} s_1 \wedge s_1 \neq s_1'}{s_1' \xrightarrow{r} s_1 \cup s_2 \wedge \text{Delete } s_1' \xrightarrow{r} s_1}$$

Figure 10-45. Rule Union_5

Rule Union6, as shown in Figure 10-46, indicates that from $s_2' \xrightarrow{r} s_2 \wedge s_2 \neq s_2'$, we (1) infer $s_2' \xrightarrow{r} s_1 \cup s_2$ and (2) delete the $s_2' \xrightarrow{r} s_2$ transition.

$$\frac{s_2' \xrightarrow{r} s_2 \wedge s_2 \neq s_2'}{s_2' \xrightarrow{r} s_1 \cup s_2 \wedge \text{Delete } s_2' \xrightarrow{r} s_2}$$

Figure 10-46. Rule Union_6

Rule Union7, as shown in Figure 10-47, indicates that from $s_1' \xrightarrow{r} s_1'' \wedge \text{Equivalent}(s_1'', s_1) \wedge s_1' = s_1''$, we (1) infer $s_1 \cup s_2 \xrightarrow{r} s_1 \cup s_2$ and (2) delete the $s_1' \xrightarrow{r} s_1''$ transition.

$$\frac{s_1' \xrightarrow{r} s_1'' \bigwedge \text{Equivalent}(s_1'', s_1) \bigwedge s_1' == s_1''}{s_1 \cup s_2 \xrightarrow{r} s_1 \cup s_2 \quad \bigwedge \quad \text{Delete} \quad s_1' \xrightarrow{r} s_1''}$$

Figure 10-47. Rule Union$_7$

Rule Union$_8$, as shown in Figure 10-48, indicates that from $s_2' \xrightarrow{r} s_2'' \bigwedge \text{Equivalent}(s_2'', s_2) \bigwedge s_2' == s_2''$, we (1) infer $s_1 \cup s_2 \xrightarrow{r} s_1 \cup s_2$ and (2) delete the $s_2' \xrightarrow{r} s_2''$ transition.

$$\frac{s_2' \xrightarrow{r} s_2'' \bigwedge \text{Equivalent}(s_2'', s_2) \bigwedge s_2' == s_2''}{s_1 \cup s_2 \xrightarrow{r} s_1 \cup s_2 \quad \bigwedge \quad \text{Delete} \quad s_2' \xrightarrow{r} s_2''}$$

Figure 10-48. Rule Union$_8$

Rule Union$_9$, as shown in Figure 10-49, indicates that from $s_1' \xrightarrow{r} s_1'' \bigwedge \text{Equivalent}(s_1'', s_1) \bigwedge s_1' \neq\neq s_1''$, we (1) infer $s_1' \xrightarrow{r} s_1 \cup s_2$ and (2) delete the $s_1' \xrightarrow{r} s_1''$ transition.

$$\frac{s_1' \xrightarrow{r} s_1'' \bigwedge \text{Equivalent}(s_1'', s_1) \bigwedge s_1' \neq\neq s_1''}{s_1' \xrightarrow{r} s_1 \cup s_2 \quad \bigwedge \quad \text{Delete} \ s_1' \xrightarrow{r} s_1''}$$

Figure 10-49. Rule Union$_9$

Rule Union_{10}, as shown in Figure 10-50, indicates that from $s_2' \xrightarrow{r} s_2'' \bigwedge \text{Equivalent}(s_2'', s_2) \bigwedge s_2' \neq s_2''$, we (1) infer $s_2' \xrightarrow{r} s_1 \cup s_2$ and (2) delete the $s_2' \xrightarrow{r} s_2''$ transition.

$$\frac{s_2' \xrightarrow{r} s_2'' \bigwedge \text{Equivalent}(s_2'', s_2) \bigwedge s_2' \neq s_2''}{s_2' \xrightarrow{r} s_1 \cup s_2 \quad \bigwedge \quad \text{Delete } s_2' \xrightarrow{r} s_2''}$$

Figure 10-50. Rule Union_{10}

Let us use four examples to illustrate the rule of Union. In the first example, we define the state expression "s_{211}".as "$r_{211} \bullet r_{212} \bullet s_{213}$" and state expression "$s_{221}$".as "$r_{221} \bullet s_{222}$". Figure 10-51 shows the diagram of the SBC state machine "SSM_{211}" which represents the state expression "s_{211}".

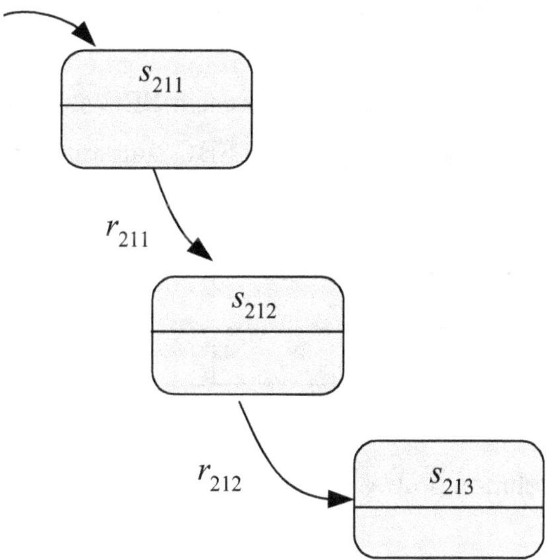

Figure 10-51. Diagram of the SBC State Machine SSM_{211}

We can also list the relationships that represent SBC state machine. Figure 10-52 shows the transition relation "$SSMR_{211}$" of the SBC state machine "SSM_{211}".

Ψ_1	R	Ψ_2
s_{211}	r_{211}	s_{212}
s_{212}	r_{212}	s_{213}

Figure 10-52. Relation $SSMR_{211}$ of the SBC State Machine SSM_{211}

Figure 10-53 shows the diagram of the SBC state machine "SSM_{221}" which represents the state expression "s_{221}".

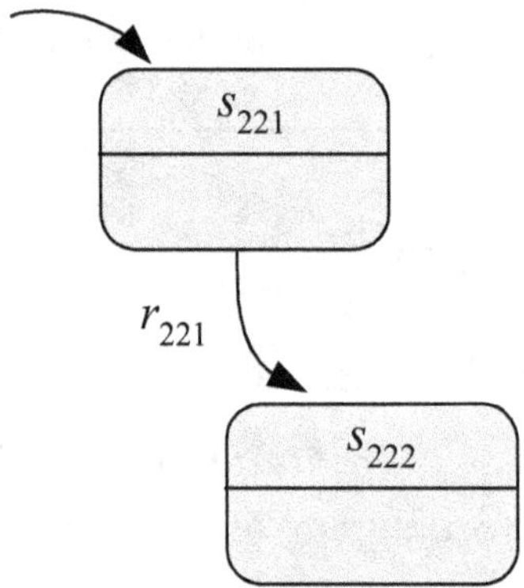

Figure 10-53. Diagram of the SBC State Machine SSM_{221}

We can also list the relationships that represent SBC state machine. Figure 10-54 shows the transition relation "$SSMR_{221}$" of the SBC state machine "SSM_{221}".

Ψ_1	R	Ψ_2
s_{221}	r_{221}	s_{222}

Figure 10-54. Relation $SSMR_{221}$ of the SBC State Machine SSM_{221}

We use the SBC state machine "SSM_{231}" to represent the state expression "$s_{211} \cup s_{221}$". According to the rule of Union, the diagram of the SBC state machine "SSM_{231}" which represents the state expression "$s_{211} \cup s_{221}$" is shown in Figure 10-55. In the state "$s_{211} \cup s_{221}$", the choice of prefixes "r_{211}" and "r_{221}" to be executed is arbitrary and fair.

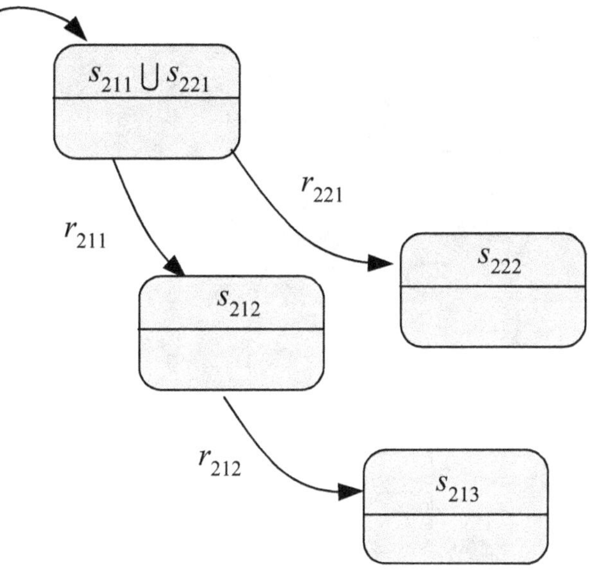

Figure 10-55. Diagram of the SBC State Machine SSM_{231}

We can also list the relationships that represent SBC state machine. Figure 10-56 shows the transition relation "$SSMR_{231}$" of the SBC state machine "SSM_{231}".

Ψ_1	R	Ψ_2
$s_{211} \cup s_{221}$	r_{211}	s_{212}
s_{212}	r_{212}	s_{213}
$s_{211} \cup s_{221}$	r_{221}	s_{222}

Figure 10-56. Relation $SSMR_{231}$ of the SBC State Machine SSM_{231}

In the second example, we define the state expression "s_{241}" as "**fix**(X_{241}=r_{241}•r_{242}•r_{243}•X_{241})" and state expression "s_{251}".as "r_{251}•s_{252}". Figure 10-57 shows the diagram of the SBC state machine "SSM_{241}" which represents the state expression "s_{241}".

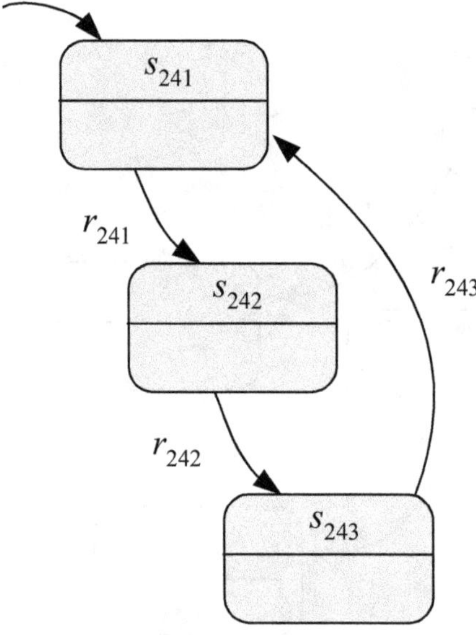

Figure 10-57. Diagram of the SBC State Machine SSM_{241}

We can also list the relationships that represent SBC state machine. Figure 10-58 shows the transition relation "$SSMR_{241}$" of the SBC state machine "SSM_{241}".

Ψ_1	R	Ψ_2
s_{241}	r_{241}	s_{242}
s_{242}	r_{242}	s_{243}
s_{243}	r_{243}	s_{241}

Figure 10-58. Relation $SSMR_{241}$ of the SBC State Machine SSM_{241}

Figure 10-59 shows the diagram of the SBC state machine "SSM_{251}" which represents the state expression "s_{251}".

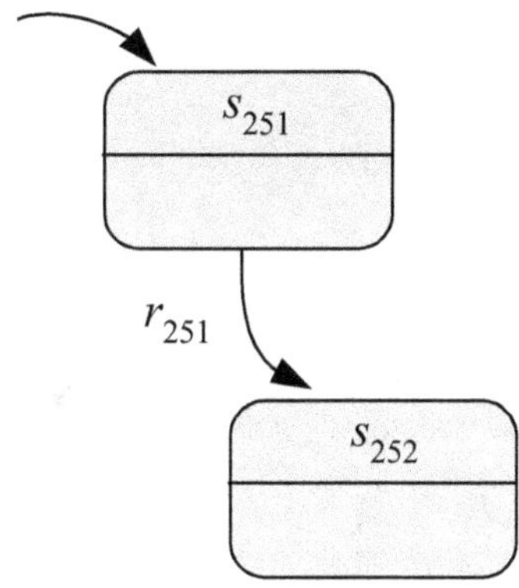

Figure 10-59. Diagram of the SBC State Machine SSM_{251}

We can also list the relationships that represent SBC state machine. Figure 10-60 shows the transition relation "$SSMR_{251}$" of the SBC state machine "SSM_{251}".

Ψ_1	R	Ψ_2
s_{251}	r_{251}	s_{252}

Figure 10-60. Relation $SSMR_{251}$ of the SBC State Machine SSM_{251}

We use the SBC state machine "SSM_{261}" to represent the state expression "$s_{241} \cup s_{251}$". According to the rule of Union, the diagram of the SBC state machine "SSM_{261}" which represents the state expression "$s_{241} \cup s_{251}$" is shown in Figure 10-61. In the state "$s_{241} \cup s_{251}$", the choice of prefixes "r_{211}" and "r_{221}" to be executed is arbitrary and fair.

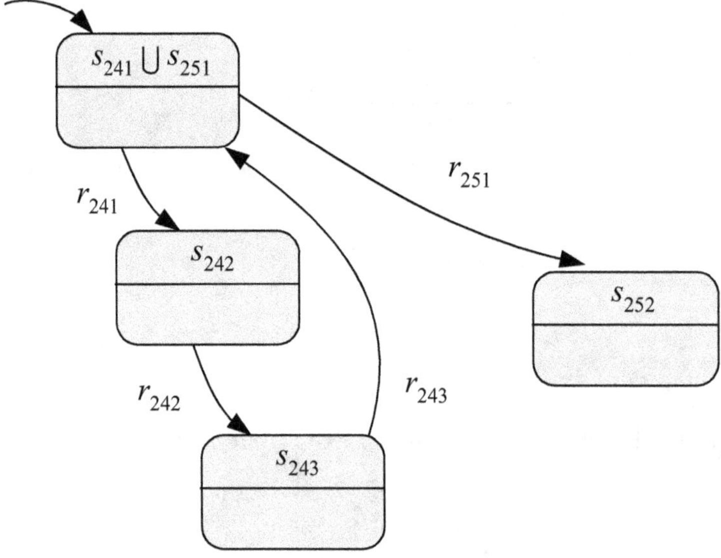

Figure 10-61. Diagram of the SBC State Machine SSM_{261}

We can also list the relationships that represent SBC state machine. Figure 10-62 shows the transition relation "$SSMR_{261}$" of the SBC state machine "SSM_{261}".

Ψ_1	R	Ψ_2
$s_{241} \cup s_{251}$	r_{241}	s_{242}
s_{242}	r_{242}	s_{243}
s_{243}	r_{243}	$s_{241} \cup s_{251}$
$s_{241} \cup s_{251}$	r_{251}	s_{252}

Figure 10-62. Relation $SSMR_{261}$ of the SBC State Machine SSM_{261}

In the third example, we define the state expression "s_{271}" as "**fix**($X_{271}=r_{271}\bullet r_{272}\bullet r_{273}\bullet X_{271}$)" and state expression "s_{281}".as "**fix**($X_{281}=r_{281}\bullet r_{282}\bullet X_{281}$)". Figure 10-63 shows the diagram of the SBC state machine "SSM_{271}" which represents the state expression "s_{271}".

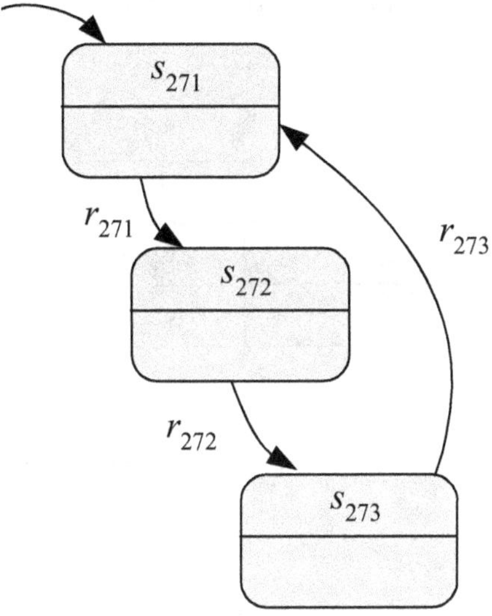

Figure 10-63. Diagram of the SBC State Machine SSM_{271}

We can also list the relationships that represent SBC state machine. Figure 10-64 shows the transition relation "$SSMR_{271}$" of the SBC state machine "SSM_{271}".

Ψ_1	R	Ψ_2
s_{271}	r_{271}	s_{272}
s_{272}	r_{272}	s_{273}
s_{273}	r_{273}	s_{271}

Figure 10-64. Relation $SSMR_{271}$ of the SBC State Machine SSM_{271}

Figure 10-65 shows the diagram of the SBC state machine "SSM_{281}" which represents the state expression "s_{281}".

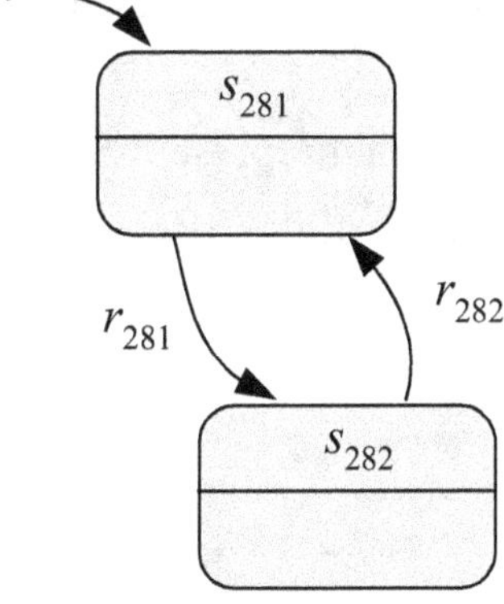

Figure 10-65. Diagram of the SBC State Machine SSM_{281}

We can also list the relationships that represent SBC state machine. Figure 10-66 shows the transition relation “$SSMR_{281}$“ of the SBC state machine “SSM_{281}“.

Ψ_1	R	Ψ_2
s_{281}	r_{281}	s_{282}
s_{282}	r_{282}	s_{281}

Figure 10-66. Relation $SSMR_{281}$ of the SBC State Machine SSM_{281}

We use the SBC state machine “SSM_{291}“ to represent the state expression “$s_{271} \cup s_{281}$”. According to the rule of Union, the diagram of the SBC state machine “SSM_{291}“ which represents the state expression “$s_{271} \cup s_{281}$” is shown in Figure 10-67.

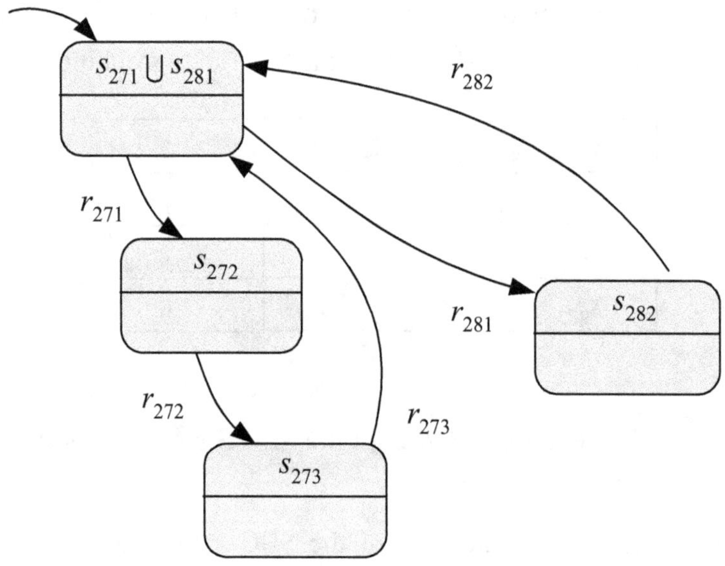

Figure 10-67. Diagram of the SBC State Machine SSM_{291}

We can also list the relationships that represent SBC state machine. Figure 10-68 shows the transition relation “$SSMR_{291}$“ of the SBC state machine “SSM_{291}“.

Ψ_1	R	Ψ_2
$s_{271} \cup s_{281}$	r_{271}	s_{272}
s_{272}	r_{272}	s_{273}
s_{273}	r_{273}	$s_{271} \cup s_{281}$
$s_{271} \cup s_{281}$	r_{281}	s_{282}
s_{282}	r_{282}	$s_{271} \cup s_{281}$

Figure 10-68. Relation $SSMR_{291}$ of the SBC State Machine SSM_{291}

In the fourth example, we define the state expression "s_{301}" as "$r_{301}\bullet\mathbf{fix}(X_{302}=r_{302}\bullet r_{303}\bullet r_{301}\bullet X_{302})$" and state expression "$s_{311}$".as "$r_{311}\bullet\mathbf{fix}(X_{312}=r_{312}\bullet r_{311}\bullet X_{312})$". Figure 10-69 shows the diagram of the SBC state machine "SSM_{301}" which represents the state expression "s_{301}".

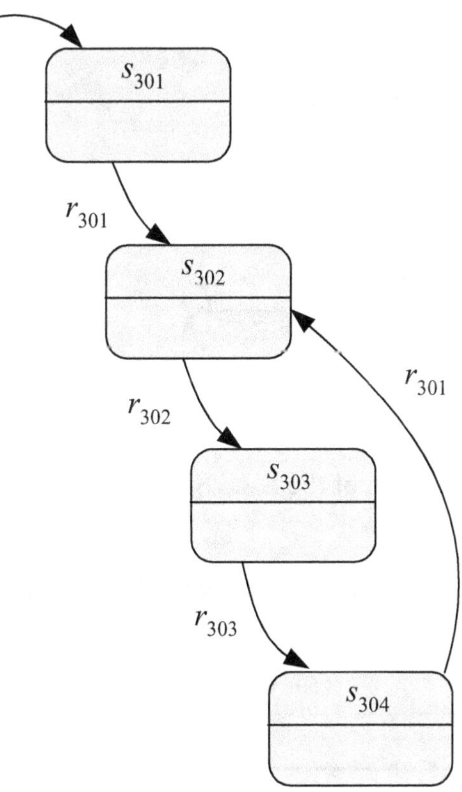

Figure 10-69. Diagram of the SBC State Machine SSM_{301}

We can also list the relationships that represent SBC state machine. Figure 10-70 shows the transition relation "$SSMR_{301}$" of the SBC state machine "SSM_{301}".

Ψ_1	R	Ψ_2
s_{301}	r_{301}	s_{302}
s_{302}	r_{302}	s_{303}
s_{303}	r_{303}	s_{304}
s_{304}	r_{301}	s_{302}

Figure 10-70. Relation $SSMR_{301}$ of the SBC State Machine SSM_{301}

Figure 10-71 shows the diagram of the SBC state machine "SSM_{311}" which represents the state expression "s_{311}".

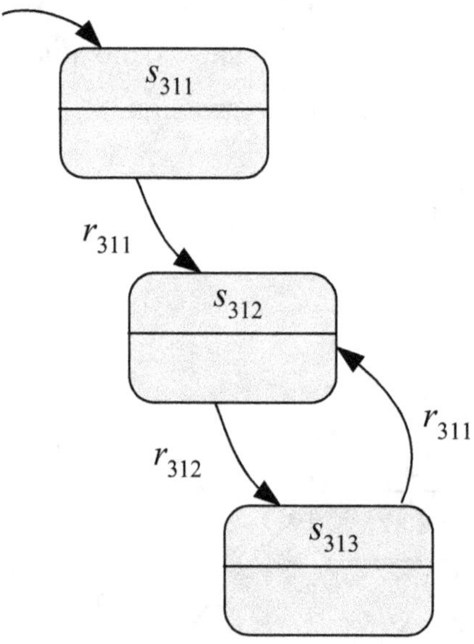

Figure 10-71. Diagram of the SBC State Machine SSM_{311}

We can also list the relationships that represent SBC state machine. Figure 10-72 shows the transition relation "$SSMR_{311}$" of the SBC state machine "SSM_{311}".

Ψ_1	R	Ψ_2
s_{311}	r_{311}	s_{312}
s_{312}	r_{312}	s_{313}
s_{313}	r_{311}	s_{312}

Figure 10-72. Relation $SSMR_{311}$ of the SBC State Machine SSM_{311}

We use the SBC state machine "SSM_{321}" to represent the state expression "$s_{301} \cup s_{311}$". According to the rule of Union, the diagram of the SBC state machine "SSM_{321}" which represents the state expression "$s_{301} \cup s_{311}$" is shown in Figure 10-73.

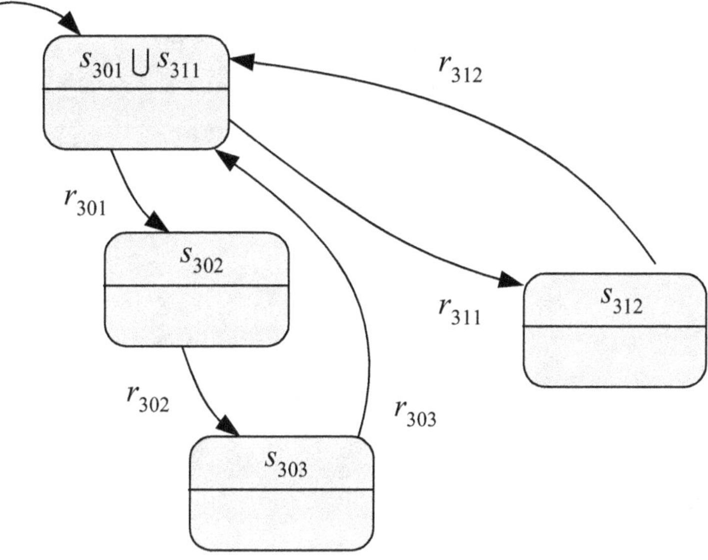

Figure 10-73. Diagram of the SBC State Machine SSM_{321}

We can also list the relationships that represent SBC state machine. Figure 10-74 shows the transition relation "$SSMR_{321}$" of the SBC state machine "SSM_{321}".

Ψ_1	R	Ψ_2
$s_{301} \cup s_{311}$	r_{301}	s_{302}
s_{302}	r_{302}	s_{303}
s_{303}	r_{303}	$s_{301} \cup s_{311}$
$s_{301} \cup s_{311}$	r_{311}	s_{312}
s_{312}	r_{312}	$s_{301} \cup s_{311}$

Figure 10-71. Relation $SSMR_{321}$ of the SBC State Machine SSM_{321}

Chapter 11: Definition of a System in SBC State Machine

Systems structure and systems behavior are the two most prominent views of a system, integrating the systems structure and systems behavior is apparently the best way to achieve an integrated whole of a system. If we are not able to integrate the systems structure and systems behavior, then there is no way that we are able to integrate the whole system. Structure-behavior coalescence (SBC) provides an elegant way to integrate the systems structure and systems behavior of a system. In other words, SBC facilitates an integrated whole of a system.

11-1 State Expression of a System

In SBC state machine, the state expression of a system is defined as $\|_{i=1,m} \text{FixIFD}_i$ and the state expression of FixIFD_i is defined as $\mathbf{fix}(X_i = \bullet_{j \in J}\, r_{ij} \bullet X_i)$. To combine them together, we summarize that in SBC state machine a system is then formally defined as "$\mathbf{fix}(X_1=r_{11}\bullet r_{12}\bullet r_{13}\bullet \ldots \bullet r_{1n}\bullet X_1) \,\|\, \mathbf{fix}(X_2=r_{21}\bullet r_{22}\bullet r_{23}\bullet \ldots \bullet r_{2n}\bullet X_2) \,\|\, \ldots \,\|\, \mathbf{fix}(X_m=r_{m1}\bullet r_{m2}\bullet r_{m3}\bullet \ldots \bullet r_{mn}\bullet X_m)$".

11-2 SBC State Machine of FixIFD

In SBC state machine, the state expression of FixIFD_i is formally defined as "$\mathbf{fix}(X_1=r_{11}\bullet r_{12}\bullet r_{13}\bullet \ldots \bullet r_{1n}\bullet X_1)$". We use the SBC state machine SSM_i to define the execution of the state expression FixIFD_i, as shown in Figure 11-1.

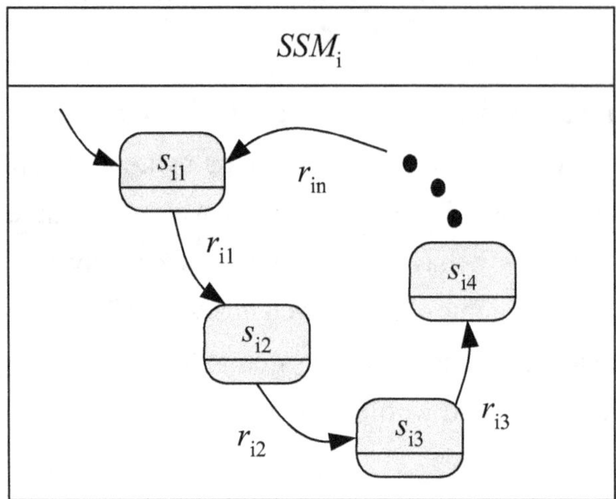

Figure 11-1. SBC State Machine "SSM_i" for the State Expression "FixIFD_i"

We can also list the relationships that represent SBC state machine. Figure 11-2 shows the transition relation "$SSMR_i$" of the SBC state machine "SSM_i".

Ψ_1	R	Ψ_2
s_{i1}	r_{i1}	s_{i2}
s_{i2}	r_{i2}	s_{i3}
s_{i3}	r_{i3}	s_{i4}
●	●	●
s_{in}	r_{in}	s_{i1}

Figure 11-2. Relation "$SSMR_i$" for the State Expression "FixIFD_i"

11-3 SBC State Machine of a System

In SBC state machine, the state expression of a system s_{system} is formally defined as "$\mathbf{fix}(X_1=g_{11}\bullet a_{12}\bullet a_{13}\bullet\ldots\bullet a_{1n}\bullet X_1) \parallel \mathbf{fix}(X_2=g_{21}\bullet a_{22}\bullet a_{23}\bullet\ldots\bullet a_{2n}\bullet X_2) \parallel \ldots \parallel \mathbf{fix}(X_m=g_{m1}\bullet a_{m2}\bullet a_{m3}\bullet\ldots\bullet a_{mn}\bullet X_m)$" or "$\text{FixIFD}_1 \parallel \text{FixIFD}_2 \parallel \ldots \parallel \text{FixIFD}_m$".

By using the SBC state machine SSM_i to define the execution of the state expression FixIFD_i, we get the SBC state machine of a system SSM_{system} defined as "$SSM_1 \| SSM_2 \| \ldots \| SSM_m$" and diagramed as shown in Figure 11-3.

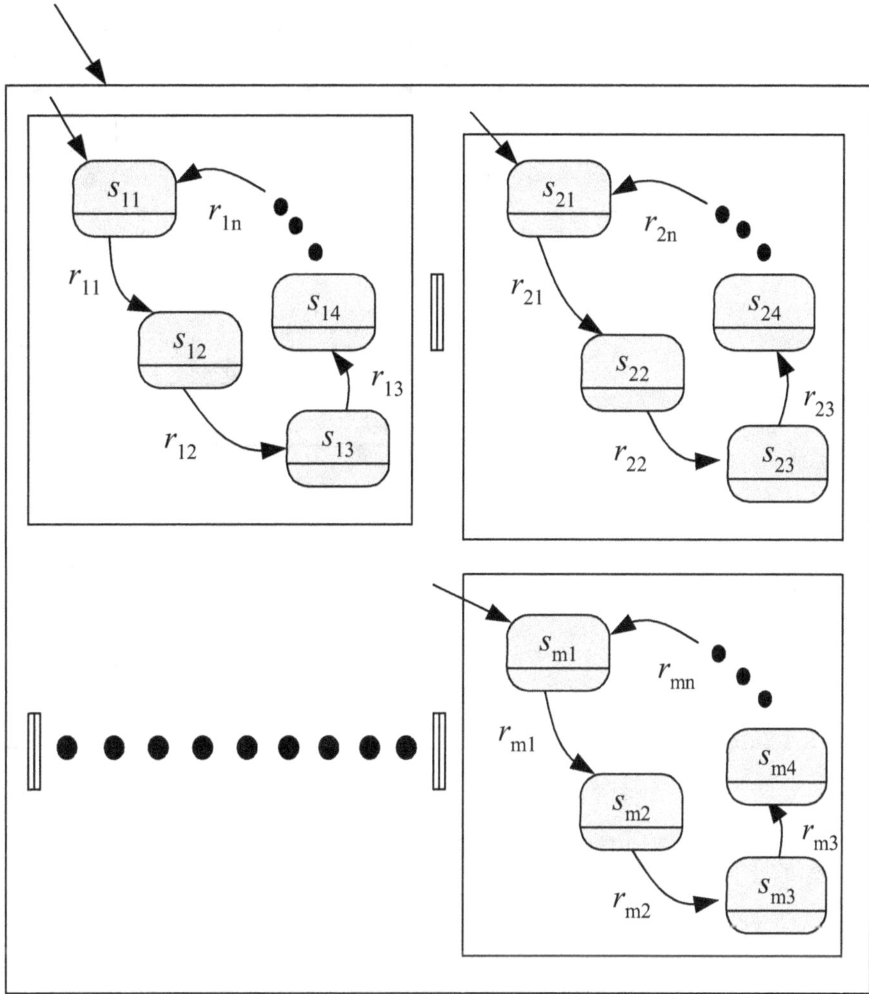

Figure 11-3. SBC State Machine SSM_{system}

We can also list the relationships that represent SBC state machine. Figure 11-4 shows the transition relation "$SSMR_{system}$" of the SBC state machine "SSM_{system}".

Ψ_1	R	Ψ_2
s_{11}	r_{11}	s_{12}
s_{12}	r_{12}	s_{13}
s_{13}	r_{13}	s_{14}
•	•	•
s_{1n}	r_{1n}	s_{11}

Ψ_1	R	Ψ_2
s_{21}	r_{21}	s_{22}
s_{22}	r_{22}	s_{23}
s_{23}	r_{23}	s_{24}
•	•	•
s_{2n}	r_{2n}	s_{21}

• • • • • • • •

Ψ_1	R	Ψ_2
s_{m1}	r_{m1}	s_{m2}
s_{m2}	r_{m2}	s_{m3}
s_{m3}	r_{m3}	s_{m4}
•	•	•
s_{mn}	r_{mn}	s_{m1}

Figure 11-4. Relation $SSMR_{\text{system}}$

Chapter 12: Basic SBC State Machines

The basic SBC state machine has the least level of complexity. Its prefix contains only pure interactions. That is, the guarded conditions are excluded from the prefix. In addition to this, no code snippet is allowed in the basic SBC state machine.

In this chapter, we first review the definitions required by the basic SBC state machine. Then, we introduce several case studies of basic SBC state machines.

12-1 Operation Call or Operation Return Signature

We formally describe the "operation call or operation return signature" as a relation $L \subseteq \Lambda \text{ X } \Theta$ where Λ is a set of "operation names" and Θ is a set of "parameter lists".

DEFINITION (OPERATION CALL OR OPERATION RETURN SIGNATURE) An Operation Call or Operation Return Signature OS = (Λ, Θ, L) consists of

. a finite set Λ of "operation names",
. a finite set Θ of "parameter lists",
. a relation $L \subseteq \Lambda \text{ X } \Theta$, and $(op, p) \in L$.

12-2 Component Operation Diagram

We formally describe the "component operation diagram" as a relation $COD \subseteq L \text{ X } \Gamma$ where L is a relation of "operation signatures" and Γ is a set of "blocks".

DEFINITION (COMPONENT OPERATION DIAGRAM) A Component Operation Diagram COD = (L, Γ, COD) consists of

- a relation L of "operation signatures",
- a finite set Γ of "blocks",
- a relation $COD \subseteq L \text{ X } \Gamma$, and $(l, b) \in COD$.

12-3 Definition of Operation-Based Value-Passing Interaction

We formally describe the "operation-based value-passing type interaction" as a relation $\Delta \subseteq N X \Xi X L X \Gamma$ where N is a set of "operation call or operation return tags" and Ξ is a set of "external environment's actors or blocks" and L is a relation of "operation call or operation return signatures" and Γ is a set of "blocks".

DEFINITION (OPERATION-BASED VALUE-PASSING INTERACTION) An Operation-Based Value-Passing Interaction OVI = $(N, \Xi, L, \Gamma, \Delta)$ consists of

- a finite set N of "operation call or operation return tags",
- a finite set Ξ of "external environment's actors or blocks",
- a relation L of " operation call or operation return signatures",
- a finite set Γ of "blocks",
- a relation $\Delta \subseteq N \text{ X } \Xi \text{ X } L \text{ X } \Gamma$, and $(n, \rho, l, b) \in \Delta$.

12-4 Definition of Basic SBC State Machine

We formally describe the "basic SBC state machine" as a transition relation $BSSMR \subseteq \Psi_1 \times \Delta \text{ v } \Psi_2$, where $(s_j, \text{a}, s_k) \in BSSMR$ is denoted by $s_j \xrightarrow{\text{a}} s_k$.

DEFINITION (BASIC SBC STATE MACHINE) A Basic SBC State Machine $BSSM = (\Psi, s_0, \Delta, BSSMR)$ consists of

- a finite non-empty set S of states,
- an initial state $s_0 \in \Psi$,
- a relation Δ of "operation-based value-passing interactions",
- a transition relation $BSSMR \subseteq \Psi_1 \times \Delta \times \Psi_2$, where $(s_j, \text{a}, s_k) \in BSSMR$ is denoted by $s_j \xrightarrow{\text{a}} s_k$.

12-5 Case Study One

In this case study, we consider an elevator door (ED). The elevator door recognizes two operations: "Command_Open" to open the door or "Command_Close" to close the door. The component operation diagram of the ED system is shown in Figure 12-1.

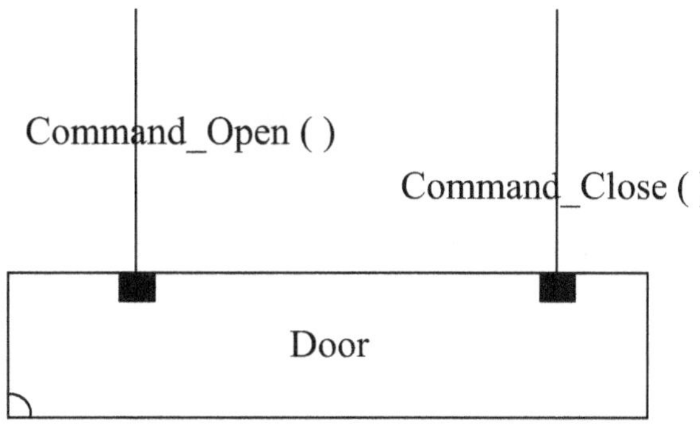

Figure 12-1. COD of the ED System

We can also list the relationships to describe the component operation diagram. Figure 12-2 shows the relation $COD \subseteq \Lambda \times \Theta \times \Gamma$ that represents the COD of the ED system.

Λ	Θ	Γ
Command_ Open		Door
Command_ Close		Door

Figure 12-2. COD Relation of the ED System

The ED system has an external actor "Passenger", which requires the "Command_Open" and "Command_Close" operations which are provided by the "Door" object. Figure 12-3 shows all those operation-based value-passing interactions that occur in the ED system.

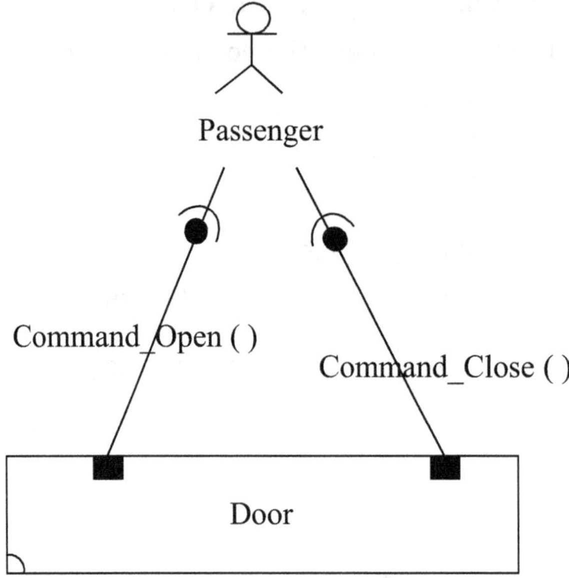

Figure 12-3. Interactions of the ED System

We can also list the relationships to describe these interactions. Figure 12-4 shows the relation $\Delta \subseteq N \times \Xi \times \Lambda \times \Theta \times \Gamma$ that represents the interactions that occur in the ED system.

Δ				
N	Ξ	Λ	Θ	Γ
a_{1201}				
CAL	Passenger	Command_ Open		Door
a_{1202}				
CAL	Passenger,	Command_ Close		Door

Figure 12-4. Interactions Relation of the ED System

Figure 12-5 describes the basic SBC state machine "$BSSM_{ED}$" of the elevator door. In the "s_{1201}" state, i.e., the door is closed, the external actor "Passenger" can interact with the "Door" block through the "Command_Open" operation call signature.

In the "s_{1202}" state, i.e., the door is opened, the external actor "Passenger" can interact with the "Door" block through the "Command_Close" operation call signature.

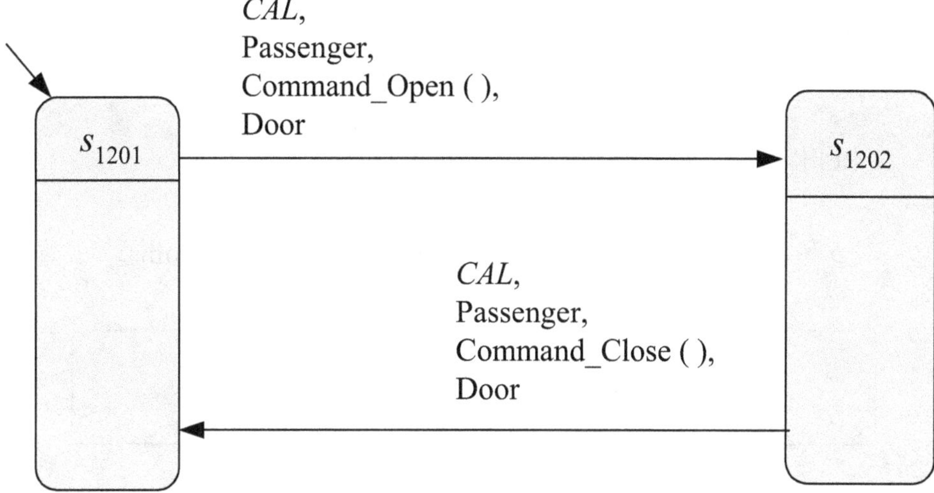

Figure 12-5. Diagram of the Basic SBC State Machine $BSSM_{\mathrm{ED}}$

We can also list the relationships that represent the basic SBC state machine. Figure 12-6 shows the transition relation "$BSSMR_{\mathrm{ED}}$" of the basic SBC state machine "$BSSM_{\mathrm{ED}}$ ".

Ψ_1	Δ					Ψ_2
	N	Ξ	Λ	Θ	Γ	
	a_{1201}					
s_{1201}	CAL	Passenger	Command_ Open		Door	s_{1202}
	a_{1202}					
s_{1202}	CAL	Passenger,	Command_ Close		Door	s_{1201}

Figure 12-6. Relation $BSSMR_{\mathrm{ED}}$ of the Basic SBC State Machine $BSSM_{\mathrm{ED}}$

12-6 Case Study Two

In this case study, "Vending Machine A" has an external actor: "Vendor" and two blocks: "Product_Store", "Coin_Store". To ensure that most vending products are available for customers to purchase, the vendor regularly refills the product store as needed. The vendor also regularly checks and refills the coin store as needed to ensure that there are deposited coins available for change. The component operation diagram of "Vending Machine A" is shown in Figure 12-7.

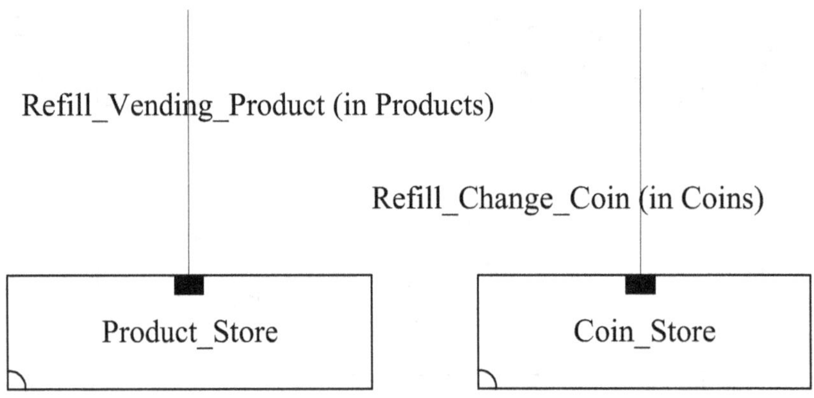

Figure 12-7. COD of "Vending Machine A"

We can also list the relationships to describe the component operation diagram. Figure 12-8 shows the relation $COD \subseteq \Lambda \times \Theta \times \Gamma$ that represents the COD of "Vending Machine A".

Λ	Θ	Γ
Refill_ Vending_ Product	in Products	Product_ Store
Refill_ Change_ Coin	in Coins	Coin_ Store

Figure 12-8. COD Relation of "Vending Machine A"

"Vending Machine A" has an external actor "Vendor", which requires the "Refill_Vending_Product (in Products)" and "Refill_Change_Coin (in Coins)" operations. The "Product_Store" block provides the "Refill_Vending_Product (in Products)" operation and the "Coin_Store" block provides "Refill_Change_Coin (in Coins)" operation. Figure 12-9 shows all those operation-based value-passing interactions that occur in "Vending Machine A".

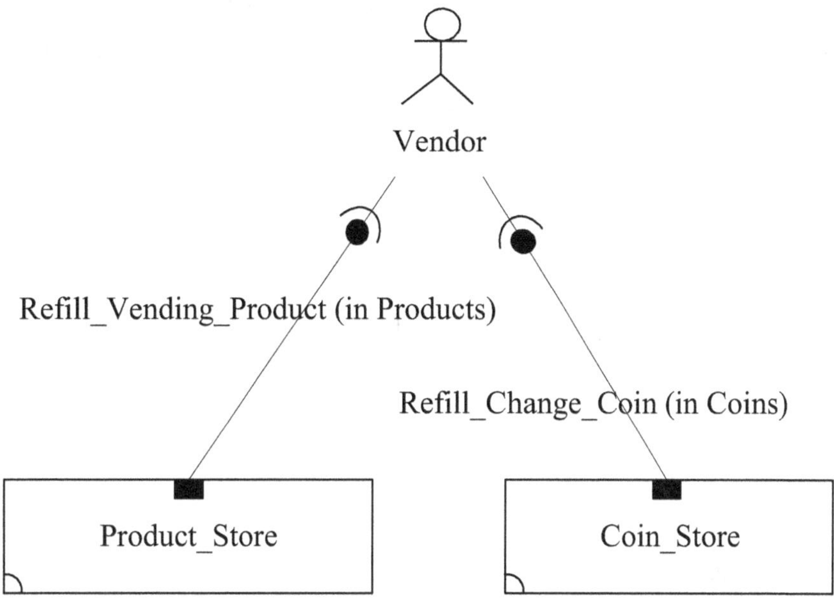

Figure 12-9. Interactions of "Vending Machine A"

We can also list the relationships to describe these interactions. Figure 12-10 shows the relation $\Delta \subseteq N \times \Xi \times \Lambda \times \Theta \times \Gamma$ that represents the interactions that occur in "Vending Machine A".

Δ				
N	Ξ	Λ	Θ	Γ
a_{1203}				
CAL	Vendor	Refill_ Vending_ Product	in Products	Product_ Store
a_{1204}				
CAL	Vendor	Refill_ Change_ Coin	in Coins	Coin_ Store

Figure 12-10. Interactions Relation of "Vending Machine A"

Figure 12-11 describes the basic SBC state machine "$BSSM_{VMA}$" of "Vending Machine A". In the "s_{1203}" state, the "Vendor" external actor can interact with the

"Product_Store" block through the "Refill_Vending_Product (in Products)" operation call signature, or interact with the "Coin_Store" block through the "Refill_Change_Coin (in Coins)" operation call signature.

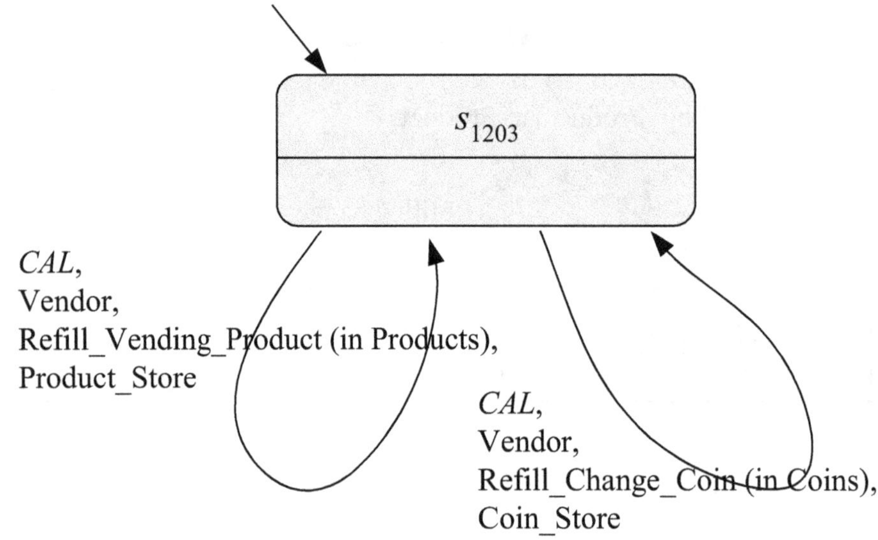

Figure 12-11. Diagram of the Basic SBC State Machine $BSSM_{\text{VMA}}$

We can also list the relationships that represent basic SBC state machine. Figure 12-12 shows the transition relation "$BSSMR_{\text{VMA}}$" of the basic SBC state machine "$BSSM_{\text{VMA}}$".

Ψ_1	Δ					Ψ_2
	N	Ξ	Λ	Θ	Γ	
	a_{1203}					
s_{1203}	*CAL*	Vendor	Refill_ Vending_ Product	in Products	Product_ Store	s_{1203}
	a_{1204}					
s_{1203}	*CAL*	Vendor	Refill_ Change_ Coin	in Coins	Coin_ Store	s_{1203}

Figure 12-12. Relation $BSSMR_{\text{VMA}}$ of the Basic SBC State Machine $BSSM_{\text{VMA}}$

12-7 Case Study Three

In this case study, “Keyboard A” has an external actor: “User” and two blocks: “Main_Keypad_A”, “Numeric_Keypad_A”. The component operation diagram of “Keyboard A” is shown in Figure 12-13.

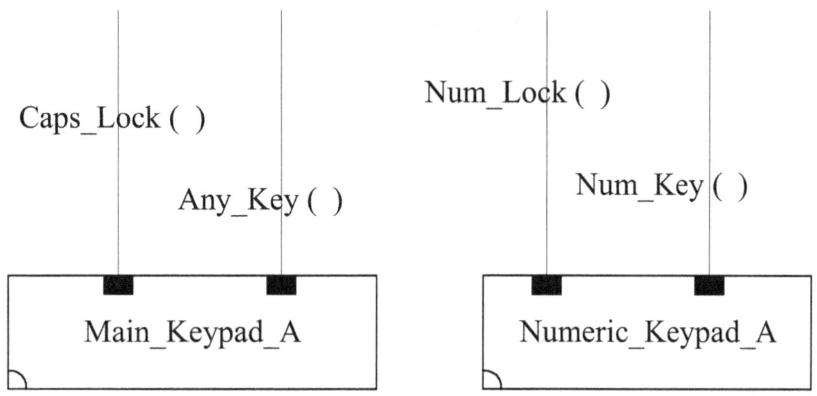

Figure 12-13. COD of "Keyboard A"

We can also list the relationships to describe the component operation diagram. Figure 12-14 shows the relation $COD \subseteq \Lambda \times \Theta \times \Gamma$ that represents the COD of “Keyboard A”.

Λ	Θ	Γ
Any_Key		Main_ Keypad_A
Caps_Lock		Main_ Keypad_A
Num_Key		Numeric_ Keypad_A
Num_Lock		Numeric_ Keypad_A

Figure 12-14. COD Relation of "Keyboard A"

In “Keyboard A”, the “User” actor requires “Caps_Lock ( )” and “Any_Key ( )” operations which are provided by the “Main_Keypad_A” block. The “Caps_Lock (

)” operation will activate or deactivate the “Caps Lock” mode. The “User” actor also requires “Num_Lock ( )” and “Num_Key ( )” operations which are provided by the “Numeric_Keypad_A” block. The “Num_Lock ( )” operation will activate or deactivate the “Num Lock” mode. Figure 12-15 shows all the operation-based value-passing interactions that occur in “Keyboard A”.

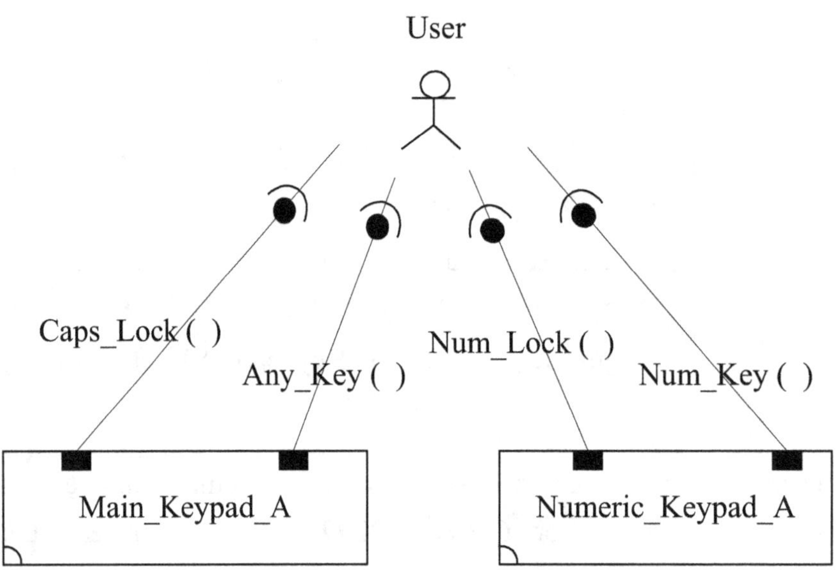

Figure 12-15. Interactions of "Keyboard A"

We can also list the relationships to describe these interactions. Figure 12-16 shows the relation $\Delta \subseteq N \text{ X } \Xi \text{ X } \Lambda \text{ X } \Theta \text{ X } \Gamma$ that represents the interactions that occur in “Keyboard A”.

Δ				
N	Ξ	Λ	Θ	Γ
a_{1211}				
CAL	User	Any_Key		Main_ Keypad_A
a_{1212}				
CAL	User	Caps_Lock		Main_ Keypad_A
a_{1213}				
CAL	User	Num_Key		Numeric_ Keypad_A
a_{1214}				
CAL	User	Num_Lock		Numeric_ Keypad_A

Figure 12-16. Interactions Relation of "Keyboard A"

Figure 12-17 describes the basic SBC state machine "$BSSM_{\mathrm{KA}}$" of "Keyboard A". The "$s_{\mathrm{keyboard\_A}}$" state has two orthogonal states: "$s_{\mathrm{main\_keypad\_A}}$" and "$s_{\mathrm{numeric\_keypad\_A}}$". The "$s_{\mathrm{main\_keypad\_A}}$" state has two composite states -- "s_{1211}" and "s_{1212}" -- depending on whether the "Caps Lock" mode is active. The "$s_{\mathrm{numeric\_keypad\_A}}$" state also has two composite states -- "s_{1213}" and "s_{1214}" -- depending on whether the "Num Lock" mode is active.

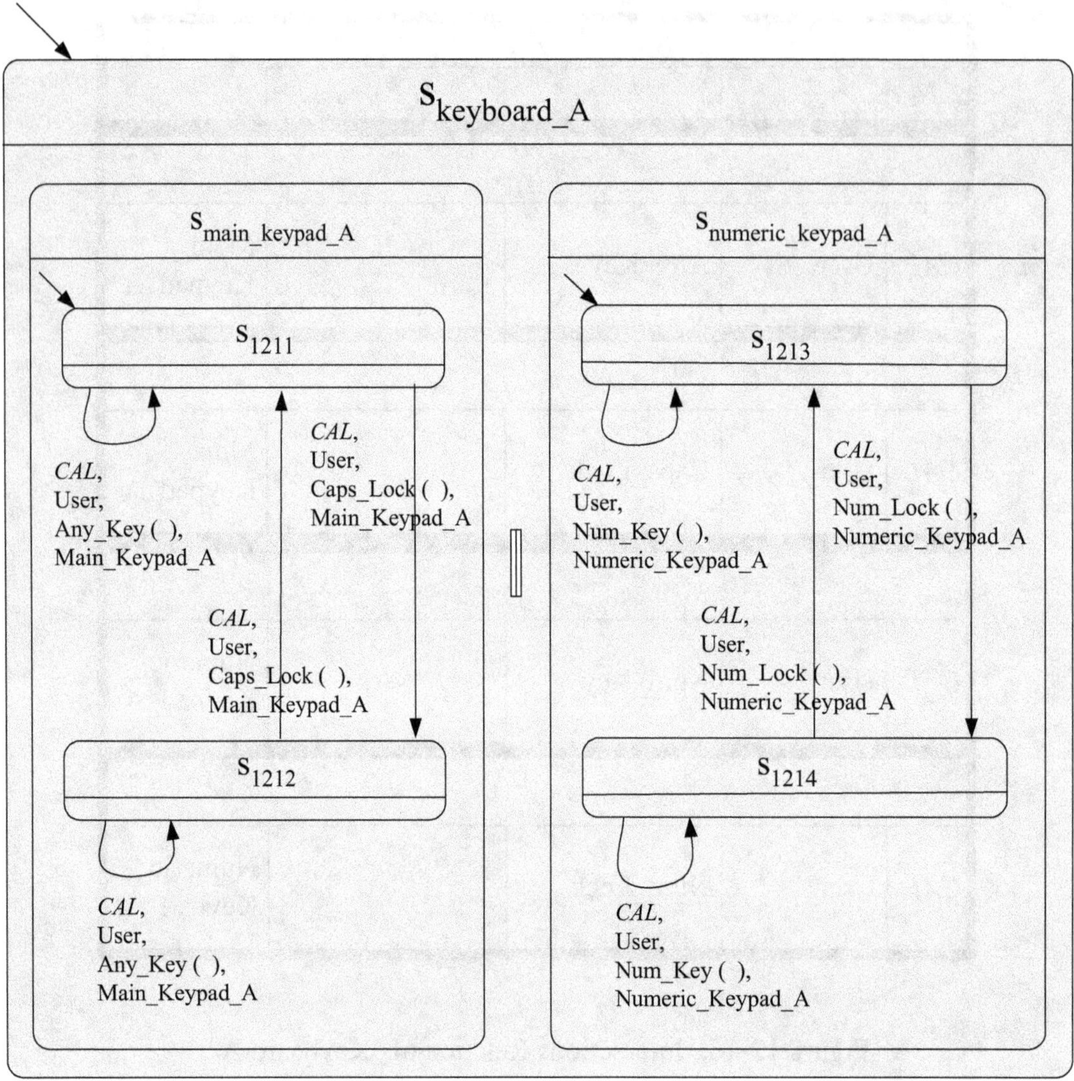

Figure 12-17. Diagram of the Basic SBC State Machine $BSSM_{KA}$

We can also list the relationships that represent the basic SBC state machine. Figure 12-18 shows the transition relation "$BSSMR_{KA}$" of the basic SBC state machine "$BSSM_{KA}$".

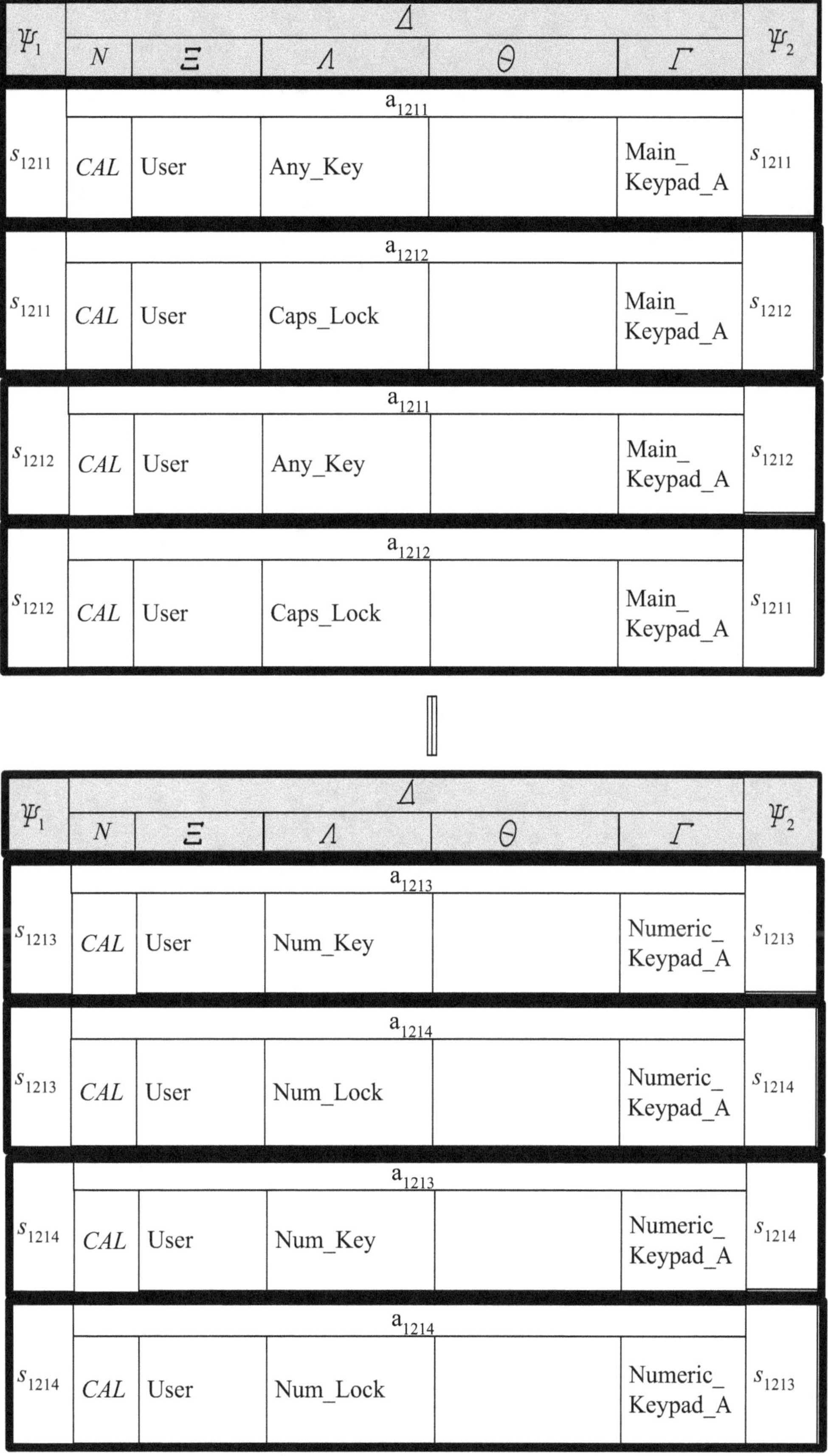

Ψ_1	Δ					Ψ_2
	N	Ξ	Λ	Θ	Γ	
	a_{1211}					
s_{1211}	*CAL*	User	Any_Key		Main_ Keypad_A	s_{1211}
	a_{1212}					
s_{1211}	*CAL*	User	Caps_Lock		Main_ Keypad_A	s_{1212}
	a_{1211}					
s_{1212}	*CAL*	User	Any_Key		Main_ Keypad_A	s_{1212}
	a_{1212}					
s_{1212}	*CAL*	User	Caps_Lock		Main_ Keypad_A	s_{1211}

$\|$

Ψ_1	Δ					Ψ_2
	N	Ξ	Λ	Θ	Γ	
	a_{1213}					
s_{1213}	*CAL*	User	Num_Key		Numeric_ Keypad_A	s_{1213}
	a_{1214}					
s_{1213}	*CAL*	User	Num_Lock		Numeric_ Keypad_A	s_{1214}
	a_{1213}					
s_{1214}	*CAL*	User	Num_Key		Numeric_ Keypad_A	s_{1214}
	a_{1214}					
s_{1214}	*CAL*	User	Num_Lock		Numeric_ Keypad_A	s_{1213}

Figure 12-18. Relation $BSSMR_{\mathrm{KA}}$ of the Basic SBC State Machine $BSSM_{\mathrm{KA}}$

Chapter 13: Intermediate SBC State Machines

The intermediate SBC state machine has the medium level of complexity. It allows guarded conditions in the prefix definition. Nonetheless, the intermediate SBC state machine still cannot contain code snippets in it.

In this chapter, we first review the definitions required by the intermediate SBC state machine. Then, we introduce some case studies of intermediate SBC state machines.

13-1 Operation Call or Operation Return Signature

We formally describe the "operation call or operation return signature" as a relation $L \subseteq \Lambda$ X Θ where Λ is a set of "operation names" and Θ is a set of "parameter lists".

DEFINITION (OPERATION CALL OR OPERATION RETURN SIGNATURE) An Operation Call or Operation Return Signature OS = (Λ, Θ, L) consists of

. a finite set Λ of "operation names",
. a finite set Θ of "parameter lists",
. a relation $L \subseteq \Lambda$ X Θ, and $(op, p) \in L$.

13-2 Component Operation Diagram

We formally describe the "component operation diagram" as a relation $COD \subseteq L$ X Γ where L is a relation of "operation signatures" and Γ is a set of "blocks".

DEFINITION (COMPONENT OPERATION DIAGRAM) A Component Operation Diagram COD = (L, Γ, COD) consists of

- a relation L of " operation signatures",
- a finite set Γ of "blocks",
- a relation $COD \subseteq L$ X Γ, and $(l, b) \in COD$.

13-3 Definition of Operation-Based Value-Passing Interaction

We formally describe the "operation-based value-passing type interaction" as a relation $\Delta \subseteq N X \Xi X L X \Gamma$ where N is a set of "operation call or operation return tags" and Ξ is a set of "external environment's actors or blocks" and L is a relation of "operation call or operation return signatures" and Γ is a set of "blocks".

DEFINITION (OPERATION-BASED VALUE-PASSING INTERACTION) An Operation-Based Value-Passing Interaction OVI = $(N, \Xi, L, \Gamma, \Delta)$ consists of

- a finite set N of "operation call or operation return tags",
- a finite set Ξ of "external environment's actors or blocks",
- a relation L of " operation call or operation return signatures",
- a finite set Γ of "blocks",
- a relation $\Delta \subseteq N \text{ X } \Xi \text{ X } L \text{ X } \Gamma$, and $(n, \rho, l, b) \in \Delta$.

13-4 Definition of Operation-Based Value-Passing Related Interaction

We formally describe the "operation-based value-passing related Interaction" as a finite set $\Omega = \{\lambda\} \cup \{NOI\} \cup \Delta$.

DEFINITION (OPERATION-BASED VALUE-PASSING RELATED INTERACTION) An Operation-Based Value-Passing Related Interaction $OVRI = (\lambda, NOI, \Delta, \Omega)$ consists of

- an internal interaction "λ",
- a non-operable interaction "*NOI* ",
- a relation Δ of "operation-based value-passing interactions",
- a finite set $\Omega = \{\lambda\} \cup \{NOI\} \cup \Delta$.

13-5 Definition of Prefix

We formally describe the "prefix" as a relation $R \subseteq C \times \Omega$ where C is a set of "optional guard conditions" and Ω is a set of "operation-based value-passing related interactions".

DEFINITION (PREFIX) A Prefix $PX = (C, \Omega, R)$ consists of

- a finite set C of optional guard conditions,
- a finite set Ω of operation-based value-passing related interactions,
- a relation $R \subseteq C \times \Omega$, and $(c, \alpha) \in R$.

13-6 Definition of Intermediate SBC State Machine

We formally describe the "intermediate SBC state machine" as a transition relation $ISSMR \subseteq \Psi_1 \times R \times \Psi_2$, where $(s_j, r, s_k) \in ISSMR$ is denoted by $s_j \xrightarrow{r} s_k$.

DEFINITION (INTERMEDIATE STATE MACHINE) An Intermediate SBC State Machine $ISSM = (\Psi, s_0, R, ISSMR)$ consists of

- a finite non-empty set Ψ of states,
- an initial state $s_0 \in \Psi$,
- a relation R of prefix,
- a transition relation $ISSMR \subseteq \Psi_1 \times R \times \Psi_2$, where $(s_j, r, s_k) \in ISSMR$ is denoted by $s_j \xrightarrow{r} s_k$.

13-7 Case Study One

In this case study, "Giraffe Chapel Cafe" is an international restaurant which has an external actor: "Customer" and three blocks: "Service_Staff", "Chinese_Cuisine_Chef", and "Mexican_Cuisine_Chef". The service staff first accepts the customer's order. According to the order, the service staff will ask the Chinese or Mexican chef to cook the corresponding food. The component operation diagram of "Giraffe Chapel Cafe" is shown in Figure 13-1.

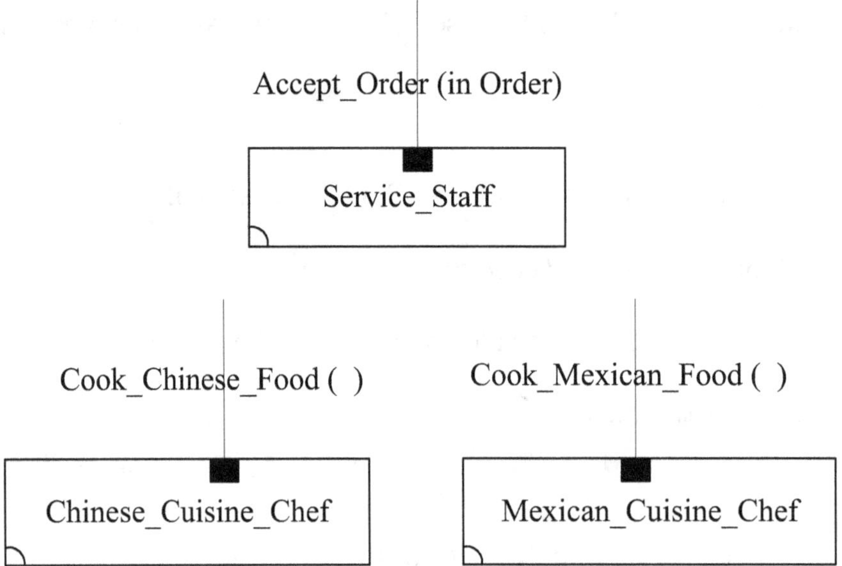

Figure 13-1. COD of "Giraffe Chapel Cafe"

We can also list the relationships to describe the component operation diagram. Figure 13-2 shows the relation $COD \subseteq \Lambda \times \Theta \times \Gamma$ that represents the COD of "Giraffe Chapel Cafe".

Λ	Θ	Γ
Accept_Order	in Order	Service_ Staff
Cook_ Chinese_Food		Chinese_ Cuisine_ Chef
Cook_ Mexican_Food		Mexican_ Cuisine_ Chef

Figure 13-2. COD Relation of "Giraffe Chapel Cafe"

In “Giraffe Chapel Cafe”, the “Customers” actor requires the “Accept_Order (in Order)” operation and the “Service_Staff” block provides it. The “Service_Staff” block requires “Cook_Chinese_Food ( )” and “Cook_Mexican_Food ( )” operations, which are provided by the “Chinese_Cuisine_Chef” and “Mexican_Cuisine_Chef” blocks, respectively. Figure 13-3 shows all these operation-based value-passing interactions that take place in the “Giraffe Chapel Cafe”.

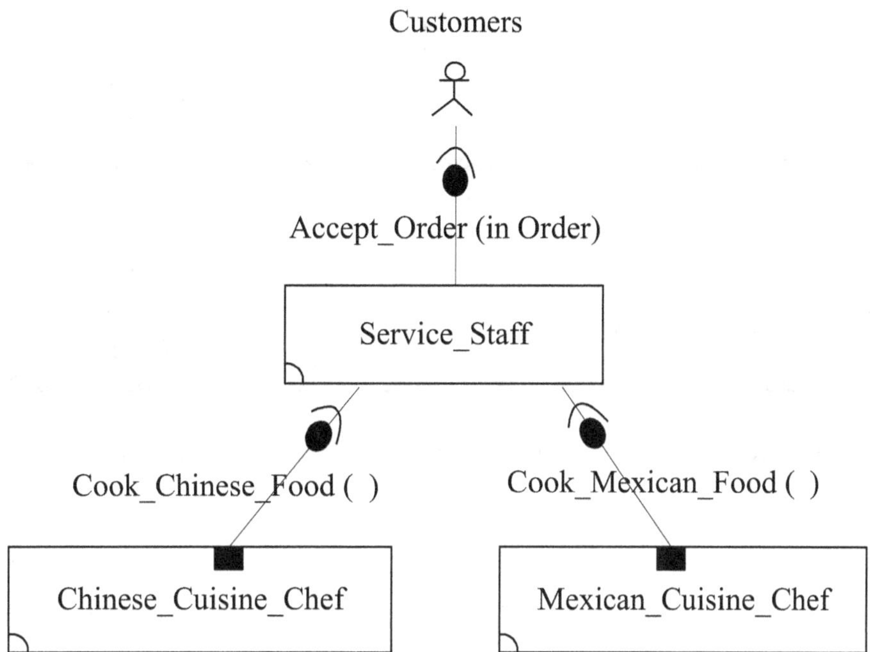

Figure 13-3. Interactions of "Giraffe Chapel Cafe"

We can also list the relationships to describe these interactions. Figure 13-4 shows the relation $\Delta \subseteq N \times \Xi \times \Lambda \times \Theta \times \Gamma$ that represents the interactions that occur in “Giraffe Chapel Cafe”.

Δ				
N	Ξ	Λ	Θ	Γ
a_{1301}				
CAL	Customers	Accept_Order	in Order	Service_ Staff
a_{1302}				
CAL	Service_ Staff	Cook_ Chinese_Food		Chinese_ Cuisine_ Chef
a_{1303}				
CAL	Service_ Staff	Cook_ Mexican_Food		Mexican_ Cuisine_ Chef

Figure 13-4. Interactions Relation of "Giraffe Chapel Cafe"

Figure 13-5 describes the intermediate SBC state machine "$ISSM_{GCC}$" of "Giraffe Chapel Cafe". In the "s_{1301}" state, the "Customer" actor can interact with the "Service_Staff" block through the "Accept_Order (in Order)" operation call signature. In the "s_{1302}" state, if the order is "Chinese_Food" then the "Service_Staff" block will interact with the "Chinese_Cuisine_Chef" block through the "Cook_Chinese_Food ( )" operation call signature; otherwise, the "Service_Staff" block will interact with the "Mexican_Cuisine_Chef" block through the "Cook_Mexican_Food ( )" operation call signature.

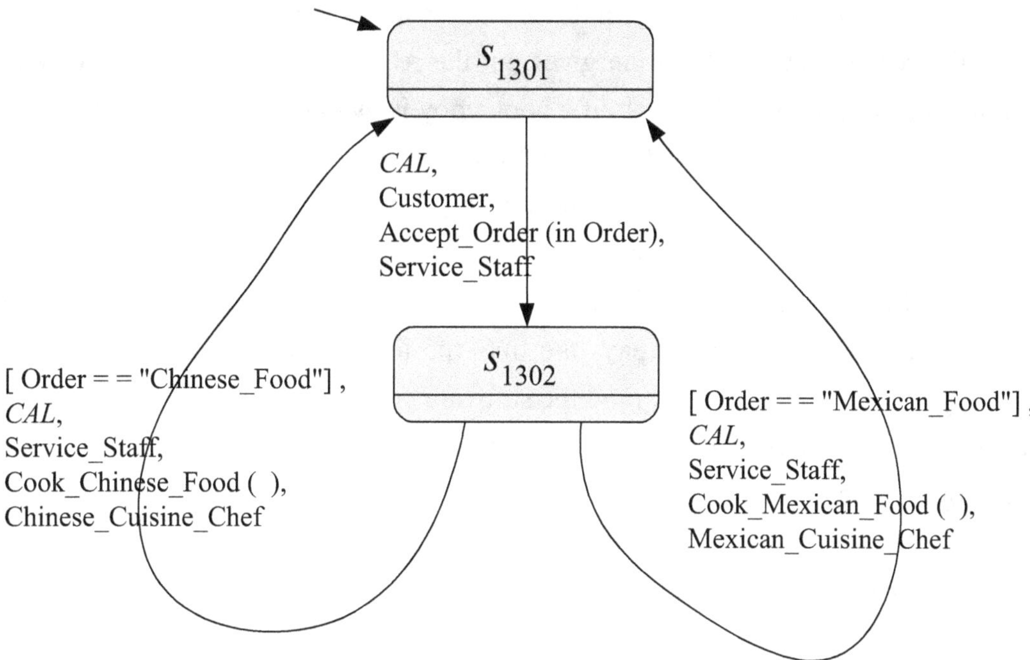

Figure 13-5. Diagram of the Intermediate SBC State Machine $ISSM_{\text{GCC}}$

We can also list the relationships that represent the intermediate SBC state machine. Figure 13-6 shows the transition relation "$ISSMR_{\text{GCC}}$" of the intermediate SBC state machine "$ISSM_{\text{GCC}}$".

Ψ_1	Δ					Ψ_2
	N	Ξ	Λ	Θ	Γ	
	a_{1301}					
s_{1301}	*CAL*	Customers	Accept_Order	in Order	Service_ Staff	s_{1302}
	a_{1302}					
s_{1302}	*CAL*	Service_ Staff	Cook_ Chinese_Food		Chinese_ Cuisine_ Chef	s_{1301}
	a_{1303}					
s_{1302}	*CAL*	Service_ Staff	Cook_ Mexican_Food		Mexican_ Cuisine_ Chef	s_{1301}

Figure 13-6. Relation $ISSMR_{\text{GCC}}$ of the Intermediate SBC State Machine $ISSM_{\text{GCC}}$

13-8 Case Study Two

In a library application, the librarian will create a reader account and card for an unregistered reader. The librarian gives out the reader card to the reader. When a librarian receives a request for book, if a book copy is available in the Book Fund, he checks out the requested book from the Book Fund to the reader; otherwise, the librarian adds the request for book to the waiting list of the Software System. When the reader returns the book copy, the librarian takes it back and returns the book to the Book Fund. He imposes the fine, if the term of the loan is exceeded, the book is lost, or is damaged. When the reader pays the fine, the librarian closes the fine. When a book copy is returned to the book fund the software system checks which requests can be satisfied and if successful informs the readers via a short message service. If the book copy is severely damaged, the librarian completes the statement of utilization, and sends the book copy to the utilizer. The component operation diagram of "the Library Application" is shown in Figure 13-7.

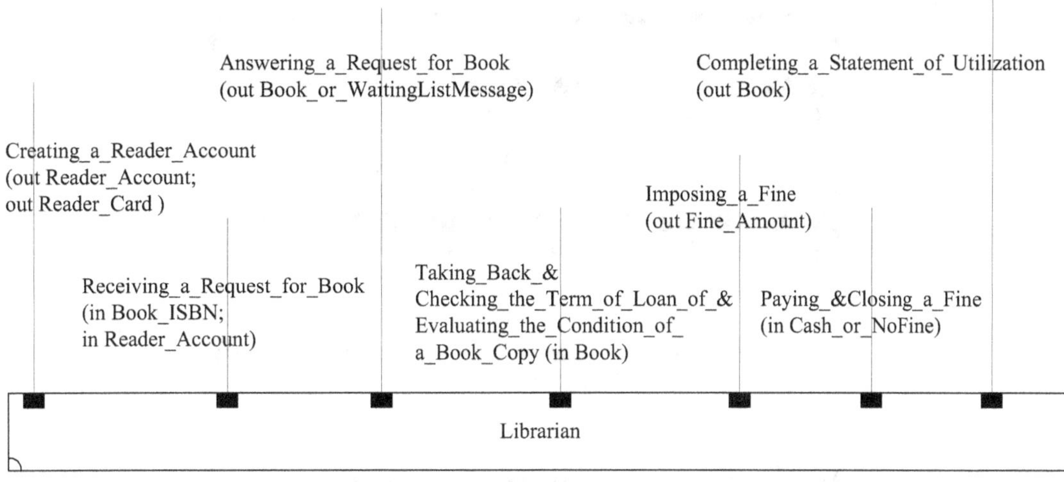

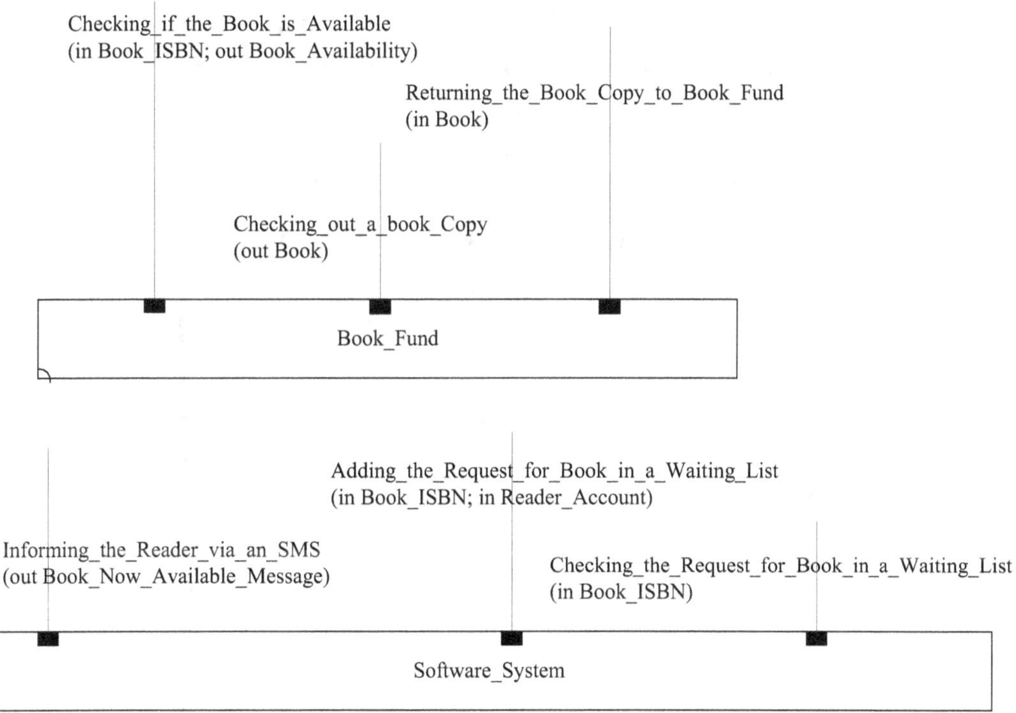

Figure 13-7. COD of "the Library Application"

We can also list the relationships to describe the component operation diagram. Figure 13-8 shows the relation $COD \subseteq \Lambda \times \Theta \times \Gamma$ that represents the COD of "the Library Application".

Λ	Θ	Γ
Creating_a_Reader_Account	out Reader_Account; out Reader_Card	Librarian
Receiving_a_Request_for_Book	in Book_ISBN; in Reader_Account	Librarian
Checking_if_the_Book_is_Available	in Book_ISBN; out Book_Availability	Book_Fund
Checking_out_a_book_Copy	out Book	Book_Fund
Adding_the_Request_for_Book_in_a_Waiting_List	in Book_ISBN; in Reader_Account	Software_System
Answering_a_Request_for_Book	out Book_or_WaitingListMessage	Librarian
Taking_Back_&Checking_the_Term_of_Loan_of_&Evaluating_the_Condition_of_a_Book_Copy	in Book	Librarian
Returning_the_Book_Copy_to_Book_Fund	in Book	Book_Fund

Figure 13-8. COD Relation of "the Library Application"

Λ	Θ	Γ
Imposing_a_Fine	out Fine_Amount	Librarian
Paying_&Closing_a_Fine	in Cash_or_NoFine	Librarian
Checking_the_Request_for_Book_in_a_Waiting_List	in Book_ISBN	Software_System
Informing_the_Reader_via_an_SMS	out Book_Now_Available_Message	Software_System
Returning_the_Book_Copy_to_Book_Fund	in Book	Book_Fund
Completing_a_Statement_of_Utilization	out Book	Librarian

Figure 13-8 (continued). COD Relation of "the Library Application"

The Library Application has two external actors: "Reader" and "Ulilizer"; three blocks: "Librarian", "Book_Fund" and "Software System". Figure 13-9 shows all the operation-based value-passing interactions that occur in "the Library Application".

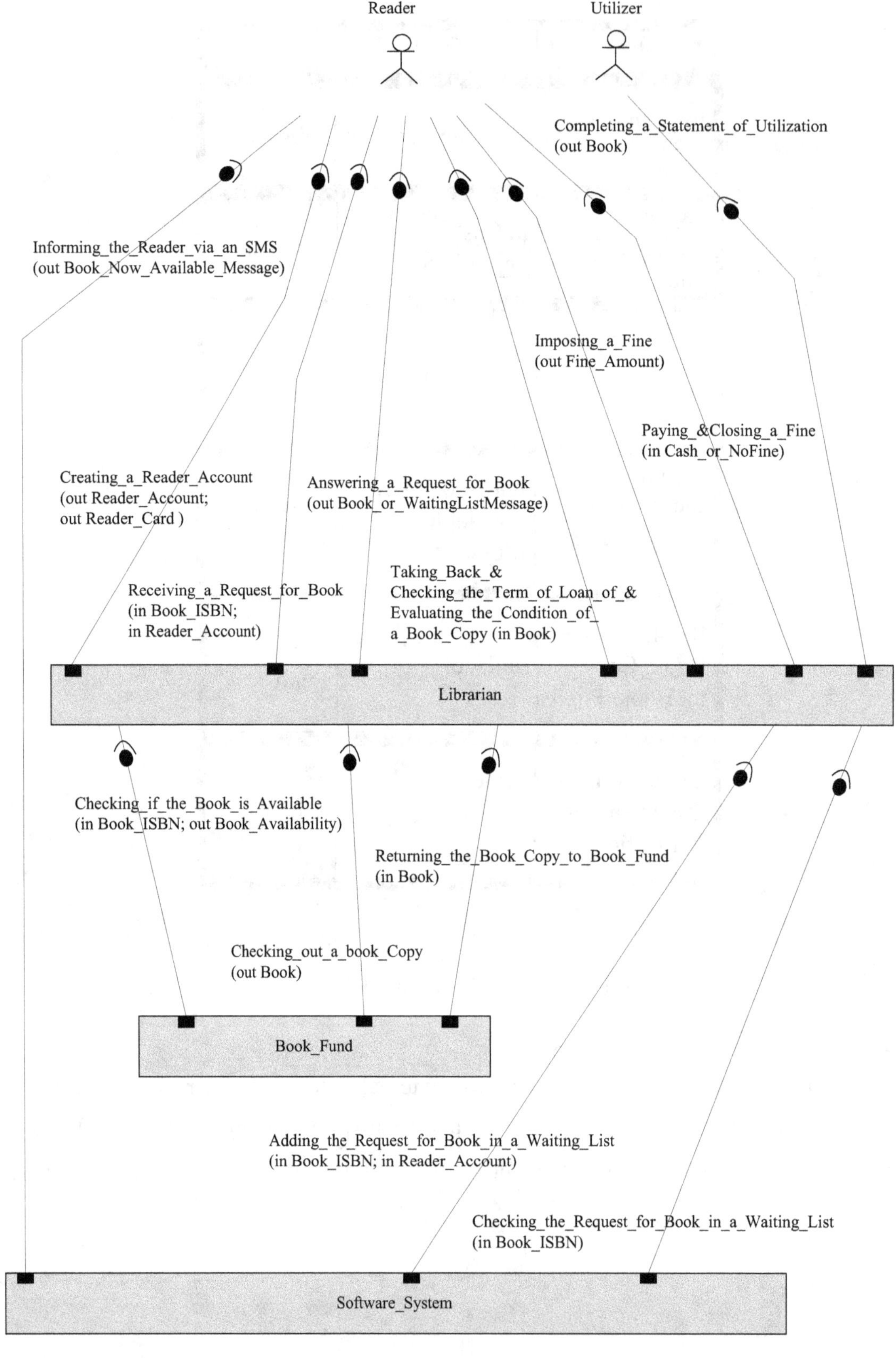

Figure 13-9. Interactions of "the Library Application"

We can also list the relationships to describe these interactions. Figure 13-10 shows the relation $\Omega = \{\lambda\} \cup \{NOI\} \cup \{N \text{ X } \Xi \text{ X } \Lambda \text{ X } \Theta \text{ X } \Gamma\}$ that represents the interactions that occur in “the Library Application”.

Ω				
N	*Ξ*	*Λ*	*Θ*	*Γ*
α_{1311}				
CAL	Reader	Creating_a_Reader_Account	out Reader_Account; out Reader_Card	Librarian
α_{1312}				
CAL	Reader	Receiving_a_Request_for_Book	in Book_ISBN; in Reader_Account	Librarian
α_{1313}				
CAL	Librarian	Checking_if_the_Book_is_Available	in Book_ISBN; out Book_Availability	Book_Fund
α_{1314}				
CAL	Librarian	Checking_out_a_book_Copy	out Book	Book_Fund
α_{1315}				
CAL	Librarian	Adding_the_Request_for_Book_in_a_Waiting_List	in Book_ISBN; in Reader_Account	Software_System
α_{1316}				
CAL	Reader	Answering_a_Request_for_Book	out Book_or_WaitingListMessage	Librarian

Figure 13-10. Interactions Relation of "the Library Application"

Ω				
N	*Ξ*	*Λ*	*Θ*	*Γ*
α_{1317}				
CAL	Reader	Taking_Back_ &Checking_ the_Term_of_ Loan_of_& Evaluating_ the_Condition_ of_a_Book_ Copy	in Book	Librarian
α_{1318}				
CAL	Librarian	Returning_the_ Book_Copy_ to_Book_Fund	in Book	Book_ Fund
α_{1319}				
CAL	Reader	Imposing_a_ Fine	out Fine_Amount	Librarian
α_{1320}				
CAL	Reader	Paying_ &Closing_a_ Fine	in Cash_ or_NoFine	Librarian
α_{1321}				
CAL	Librarian	Checking_the_ Request_for_ Book_in_a_ Waiting_List	in Book_ISBN	Software_ System
α_{1322}				
CAL	Reader	Informing_ the_Reader_ via_an_SMS	out Book_Now_ Available_ Message	Software_ System

Figure 13-10 (continued). Interactions Relation of "the Library Application"

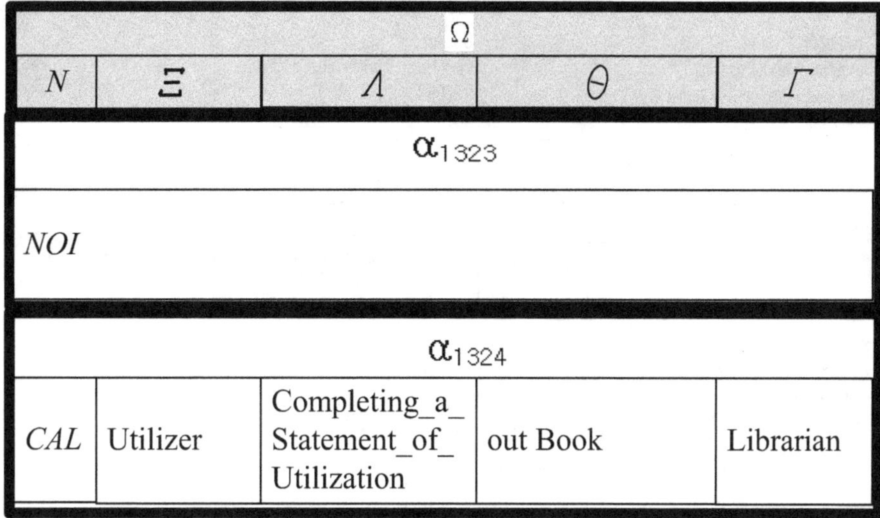

Ω				
N	Ξ	*Λ*	*Θ*	*Γ*
α_{1323}				
NOI				
α_{1324}				
CAL	Utilizer	Completing_a_ Statement_of_ Utilization	out Book	Librarian

Figure 13-10 (continued). Interactions Relation of "the Library Application"

Figure 13-11 describes the intermediate SBC state machine "$ISSM_{\text{TLA}}$" of "the Library Application".

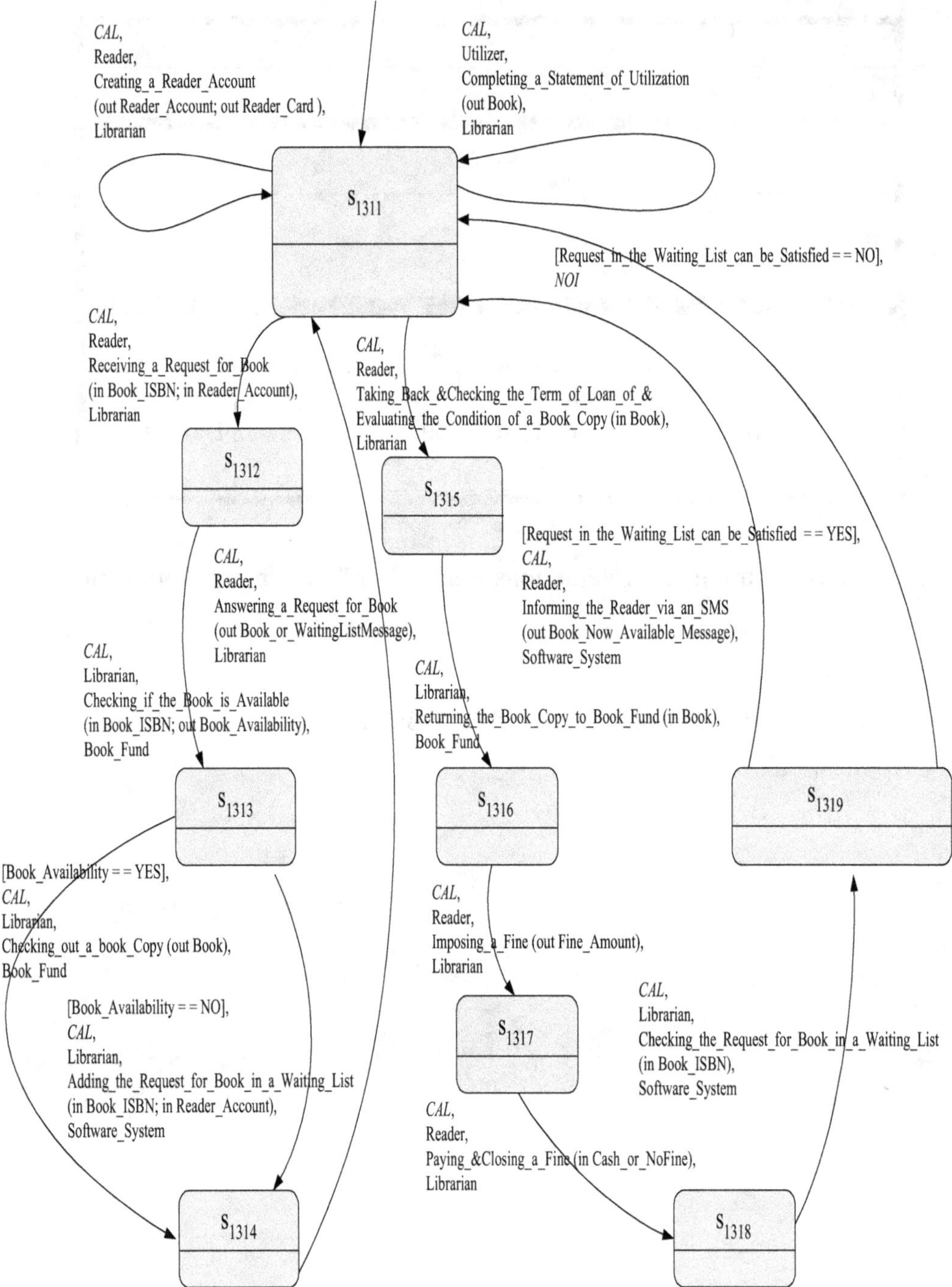

Figure 13-11. Diagram of the Intermediate SBC State Machine $ISSM_{TLA}$

We can also list the relationships that represent the intermediate SBC state machine. Figure 13-12 shows the transition relation "$ISSMR_{\text{TLA}}$" of the intermediate SBC state machine "$ISSM_{\text{TLA}}$".

<table>
<tr><th rowspan="3">Ψ_1</th><th colspan="6">R</th><th rowspan="3">Ψ_2</th></tr>
<tr><th rowspan="2">C</th><th colspan="5">Ω</th></tr>
<tr><th>N</th><th>Ξ</th><th>Λ</th><th>Θ</th><th>Γ</th></tr>
<tr><td rowspan="3">s_{1311}</td><td colspan="6">r_{1311}</td><td rowspan="3">s_{1311}</td></tr>
<tr><td rowspan="2"></td><td colspan="5">α_{1311}</td></tr>
<tr><td>*CAL*</td><td>Reader</td><td>Creating_a_ Reader_ Account</td><td>out Reader_ Account; out Reader_Card</td><td>Librarian</td></tr>
<tr><td rowspan="3">s_{1311}</td><td colspan="6">r_{1312}</td><td rowspan="3">s_{1312}</td></tr>
<tr><td rowspan="2"></td><td colspan="5">α_{1312}</td></tr>
<tr><td>*CAL*</td><td>Reader</td><td>Receiving_a_ Request_for_ Book</td><td>in Book_ISBN; in Reader_ Account</td><td>Librarian</td></tr>
</table>

Figure 13-12. Relation $ISSMR_{\text{TLA}}$ of the Intermediate SBC State Machine $ISSM_{\text{TLA}}$

Ψ_1	C	N	Ξ	Λ	Θ	Γ	Ψ_2
		R (Ω)					
		r_{1313} / α_{1313}					
s_{1312}		CAL	Librarian	Checking_if_the_Book_is_Available	in Book_ISBN; out Book_Availability	Book_Fund	s_{1313}
		r_{1314} / α_{1314}					
s_{1313}	Book_Availability == YES	CAL	Librarian	Checking_out_a_book_Copy	out Book	Book_Fund	s_{1314}
		r_{1315} / α_{1315}					
s_{1313}	Book_Availability == NO	CAL	Librarian	Adding_the_Request_for_Book_in_a_Waiting_List	in Book_ISBN; in Reader_Account	Software_System	s_{1314}
		r_{1316} / α_{1316}					
s_{1314}		CAL	Reader	Answering_a_Request_for_Book	out Book_or_WaitingListMessage	Librarian	s_{1311}
		r_{1317} / α_{1317}					
s_{1311}		CAL	Reader	Taking_Back_&Checking_the_Term_of_Loan_of_&Evaluating_the_Condition_of_a_Book_Copy	in Book	Librarian	s_{1315}
		r_{1318} / α_{1318}					
s_{1315}		CAL	Librarian	Returning_the_Book_Copy_to_Book_Fund	in Book	Book_Fund	s_{1316}

Figure 13-12 (continued). Relation $ISSMR_{TLA}$ of the Intermediate SBC State Machine $ISSM_{TLA}$

Ψ_1	C	N	Ξ	Λ	Θ	Γ	Ψ_2
		R					
		Ω					
	r_{1319}						
s_{1316}		α_{1319}					s_{1317}
		CAL	Reader	Imposing_a_Fine	out Fine_Amount	Librarian	
	r_{1320}						
s_{1317}		α_{1320}					s_{1318}
		CAL	Reader	Paying_&Closing_a_Fine	in Cash_or_NoFine	Librarian	
	r_{1321}						
s_{1318}		α_{1321}					s_{1319}
		CAL	Librarian	Checking_the_Request_for_Book_in_a_Waiting_List	in Book_ISBN	Software_System	
	r_{1322}						
s_{1319}	Request_in_the_Waiting_List_can_be_Satisfied == YES	α_{1322}					s_{1311}
		CAL	Reader	Informing_the_Reader_via_an_SMS	out Book_Now_Available_Message	Software_System	
	r_{1323}						
s_{1319}	Request_in_the_Waiting_List_can_be_Satisfied == NO	α_{1323}					s_{1311}
		NOI					
	r_{1324}						
s_{1311}		α_{1324}					s_{1311}
		CAL	Utilizer	Completing_a_Statement_of_Utilization	out Book	Librarian	

Figure 13-12 (continued). Relation $ISSMR_{\text{TLA}}$ of the Intermediate SBC State Machine $ISSM_{\text{TLA}}$

Chapter 14: Advanced SBC State Machines

The advanced SBC state machine has the highest level of complexity. It allows guarded conditions in the prefix definition. In addition, the advanced SBC state machine also contains code snippets“ in it.

In this chapter, we first review the definitions required by the advanced SBC state machine. Then, we introduce some case studies of advanced SBC state machines.

14-1 Operation Call or Operation Return Signature

We formally describe the “operation call or operation return signature” as a relation $L \subseteq \Lambda \times \Theta$ where Λ is a set of “operation names” and Θ is a set of “parameter lists”.

DEFINITION (OPERATION CALL OR OPERATION RETURN SIGNATURE) An Operation Call or Operation Return Signature OS = (Λ, Θ, L) consists of

. a finite set Λ of “operation names“,
. a finite set Θ of “parameter lists“,
. a relation $L \subseteq \Lambda \times \Theta$, and $(op, p) \in L$.

14-2 Component Operation Diagram

We formally describe the “component operation diagram” as a relation $COD \subseteq L \times \Gamma$ where L is a relation of “operation signatures” and Γ is a set of “blocks”.

DEFINITION (COMPONENT OPERATION DIAGRAM) A Component Operation Diagram COD = (L, Γ, COD) consists of

- a relation L of ”operation signatures”,
- a finite set Γ of “blocks”,
- a relation $COD \subseteq L \times \Gamma$, and $(l, b) \in COD$.

14-3 Definition of Operation-Based Value-Passing Interaction

We formally describe the "operation-based value-passing type interaction" as a relation $\Delta \subseteq N X \Xi X L X \Gamma$ where N is a set of "operation call or operation return tags" and Ξ is a set of "external environment's actors or blocks" and L is a relation of "operation call or operation return signatures" and Γ is a set of "blocks".

DEFINITION (OPERATION-BASED VALUE-PASSING INTERACTION) An Operation-Based Value-Passing Interaction OVI = $(N, \Xi, L, \Gamma, \Delta)$ consists of

- a finite set N of "operation call or operation return tags",
- a finite set Ξ of "external environment's actors or blocks",
- a relation L of " operation call or operation return signatures",
- a finite set Γ of "blocks",
- a relation $\Delta \subseteq N \text{ X } \Xi \text{ X } L \text{ X } \Gamma$, and $(n, \rho, l, b) \in \Delta$.

14-4 Definition of Operation-Based Value-Passing Related Interaction

We formally describe the "operation-based value-passing related Interaction" as a finite set $\Omega = \{\lambda\} \cup \{NOI\} \cup \Delta$.

DEFINITION (OPERATION-BASED VALUE-PASSING RELATED INTERACTION) An Operation-Based Value-Passing Related Interaction *OVRI* = $(\lambda, NOI, \Delta, \Omega)$ consists of

- an internal interaction "λ",
- a non-operable interaction "*NOI* ",
- a relation Δ of " operation-based value-passing related interactions",
- a finite set $\Omega = \{\lambda\} \cup \{NOI\} \cup \Delta$.

14-5 Definition of Prefix

We formally describe the "prefix" as a relation $R \subseteq C \times \Omega \times \Pi$ where C is a set of "optional guard conditions" and Ω is a set of "operation-based value-passing related interactions" and Π is a set of "optional code snippets".

DEFINITION (PREFIX) A Prefix $PX = (C, \Omega, \Pi, R)$ consists of

- a finite set C of optional guard conditions,
- a finite set Ω of operation-based value-passing related interactions,
- a finite set Π of optional code snippets,
- a relation $R \subseteq C \times \Omega \times \Pi$, and $(c, \alpha, \pi) \in R$.

14-6 Definition of Advanced SBC State Machine

We formally describe the "advanced SBC state machine" as a transition relation $ASSMR \subseteq \Psi_1 \times R \times \Psi_2$, where $(s_j, r, s_k) \in ASSMR$ is denoted by $s_j \xrightarrow{r} s_k$.

DEFINITION (ADVANCED SBC STATE MACHINE) An Advanced SBC State Machine $ASSM = (\Psi, (\pi_0, s_0), R, ASSMR)$ consists of

- a finite non-empty set Ψ of states,
- an optional code snippet π_0 in the initial transition, and $\pi_0 \in \Pi$,
- an initial state $s_0 \in \Psi$,
- a relation R of prefix,
- a transition relation $ASSMR \subseteq \Psi_1 \times R \times \Psi_2$, where $(s_j, r, s_k) \in ASSMR$ is denoted by $s_j \xrightarrow{r} s_k$.

14-7 Case Study One

"Keyboard B" is not much different from a normal keyboard. The only difference is that the number of keystrokes on "Keyboard B" is limited to 1000. We will introduce a key_count variable, which is initialized to 1000 and will decrease by 1 for each keystroke. The component operation diagram of "Keyboard B" is shown

Figure 14-1.

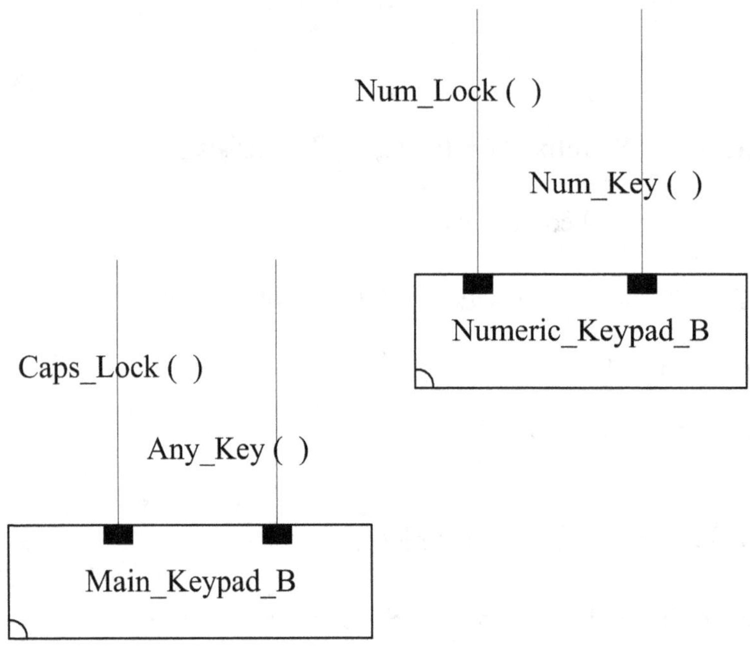

Figure 14-1. COD of "Keyboard B"

We can also list the relationships to describe the component operation diagram. Figure 14-2 shows the relation $COD \subseteq L \times \Gamma$ that represents the COD of “Keyboard B”.

L	Γ
Any_Key ( )	Main_ Keypad_A
Caps_Lock ( )	Main_ Keypad_A
Num_Key ( )	Numeric_ Keypad_A
Num_Lock ( )	Numeric_ Keypad_A

Figure 14-2. COD Relation of "Keyboard B"

In “Keyboard B”, the “User” actor requires “Caps_Lock ( )” and “Any_Key ( )” operations which are provided by the “Main_Keypad_B” block. The “User” actor also requires “Num_Lock ( )” and “Num_Key ( )” operations which are provided by the “Numeric_Keypad_B” block. Figure 14-3 shows all the operation-based value-passing interactions that occur in “Keyboard B”.

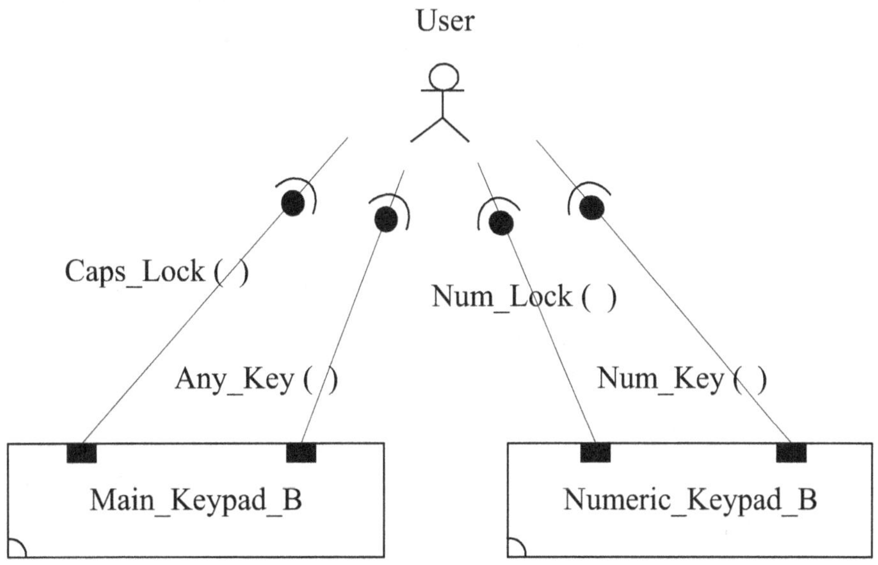

Figure 14-3. Interactions of "Keyboard B"

We can also list the relationships to describe these interactions. Figure 14-4 shows the relation $\Delta \subseteq N \times \Xi \times L \times \Gamma$ that represents the interactions that occur in "Keyboard B".

N	Ξ	L	Γ
		a_{1401}	
CAL	User	Any_Key ( )	Main_ Keypad_A
		a_{1402}	
CAL	User	Caps_Lock ( )	Main_ Keypad_A
		a_{1403}	
CAL	User	Num_Key ( )	Numeric_ Keypad_A
		a_{1404}	
CAL	User	Num_Lock ( )	Numeric_ Keypad_A

Figure 14-4. Interactions Relation of "Keyboard B"

Figure 14-5 describes the advanced SBC state machine "$ASSM_{KB}$" of "Keyboard B". The "$s_{keyboard\_B}$" state has two orthogonal states: "$s_{main\_keypad\_B}$" and "$s_{numeric\_keypad\_B}$". The "$s_{main\_keypad\_B}$" state has two composite states -- "s_{1401}" and "s_{1402}" -- depending on whether the "Caps Lock" mode is active. The "$s_{numeric\_keypad\_B}$" state also has two composite states -- "s_{1403}" and "s_{1404}" -- depending on whether the "Num Lock" mode is active.

Code snippets can be attached to the initial transition. Code snippets can also be attached to the prefix of the transitions. The advanced SBC state machine "$ASSM_{KB}$" of "Keyboard B" shows that the key_count variable is initialized to 1000 and decreases by 1 for each keystroke.

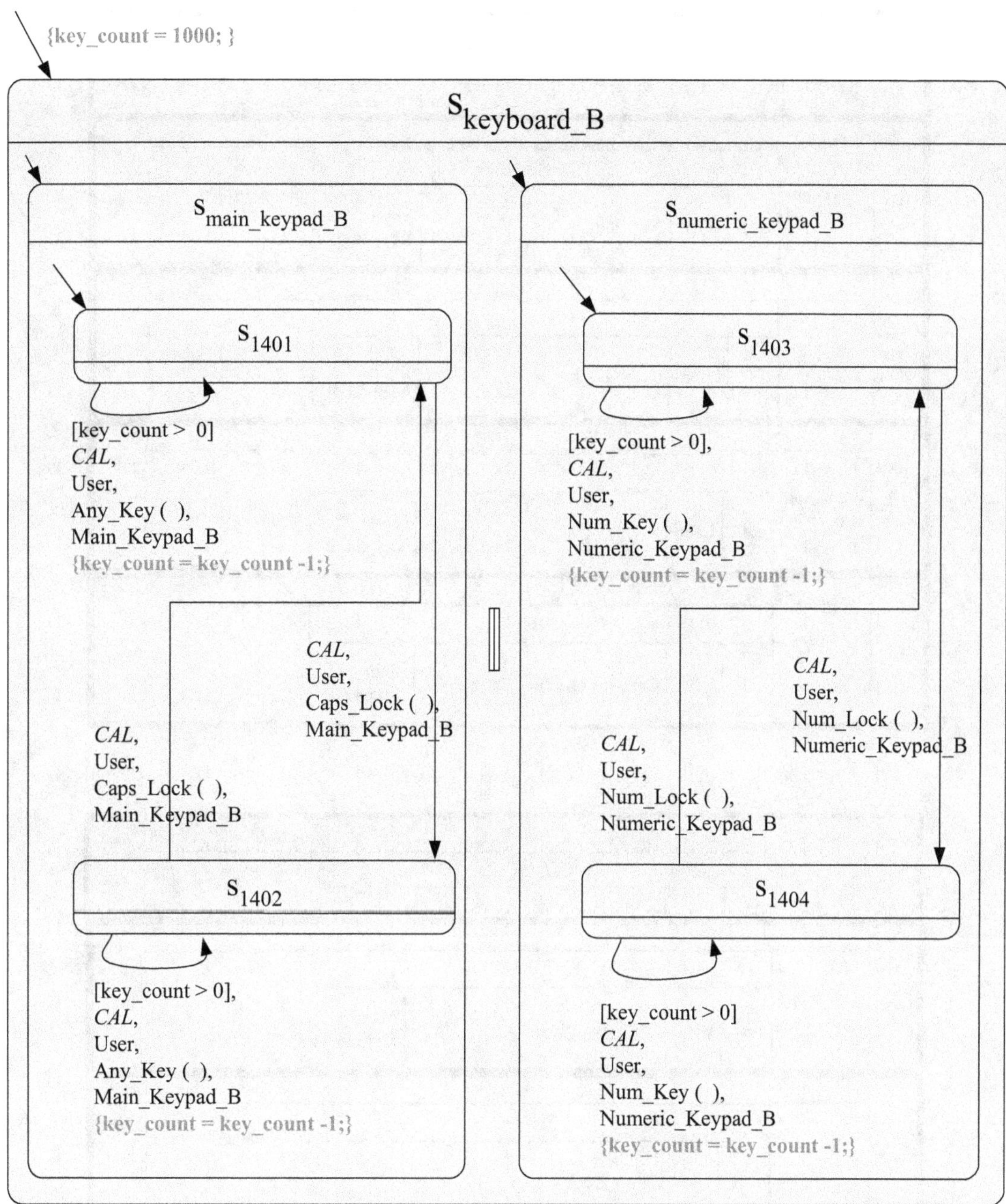

Figure 14-5. Diagram of the Advanced SBC State Machine $ASSM_{KB}$

We can also list the relationships that represent the advanced SBC state machine. Figure 14-6 shows the transition relation "$ASSMR_{KB}$" of the advanced SBC state machine "$ASSM_{KB}$".

$s_{1401} \parallel s_{1403}$
{key_count = 1000;}

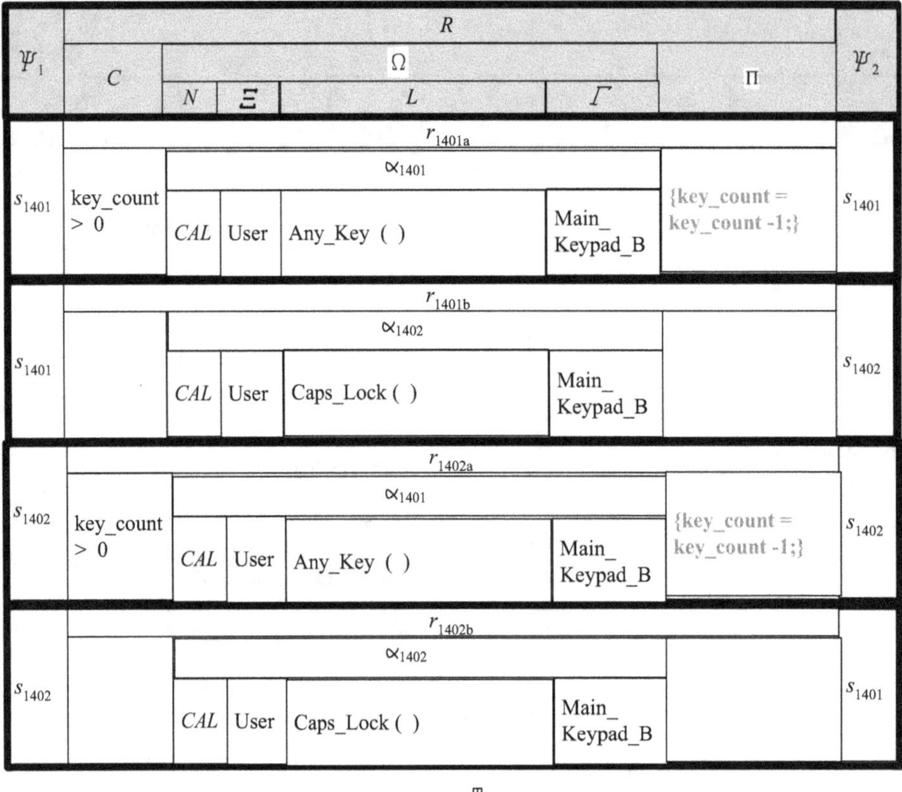

Ψ_1	R						Ψ_2
	C	Ω				Π	
		N	Ξ	L	Γ		
	r_{1401a}						
		$\propto_{1401}$					
s_{1401}	key_count > 0	CAL	User	Any_Key ( )	Main_ Keypad_B	{key_count = key_count -1;}	s_{1401}
	r_{1401b}						
		$\propto_{1402}$					
s_{1401}		CAL	User	Caps_Lock ( )	Main_ Keypad_B		s_{1402}
	r_{1402a}						
		$\propto_{1401}$					
s_{1402}	key_count > 0	CAL	User	Any_Key ( )	Main_ Keypad_B	{key_count = key_count -1;}	s_{1402}
	r_{1402b}						
		$\propto_{1402}$					
s_{1402}		CAL	User	Caps_Lock ( )	Main_ Keypad_B		s_{1401}

$\parallel$

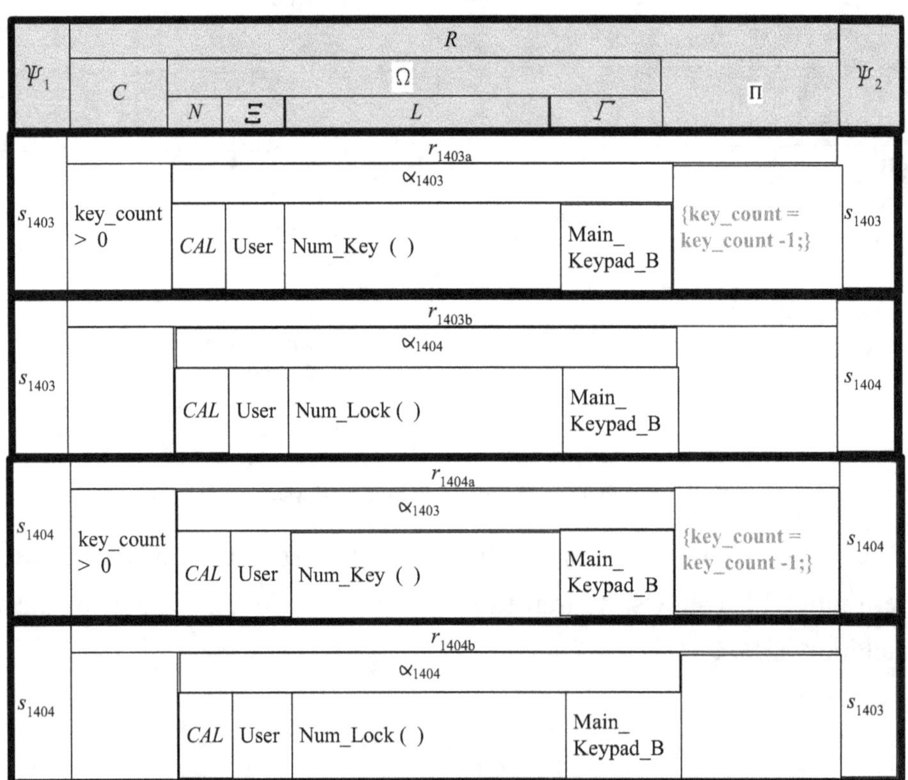

Ψ_1	R						Ψ_2
	C	Ω				Π	
		N	Ξ	L	Γ		
	r_{1403a}						
		$\propto_{1403}$					
s_{1403}	key_count > 0	CAL	User	Num_Key ( )	Main_ Keypad_B	{key_count = key_count -1;}	s_{1403}
	r_{1403b}						
		$\propto_{1404}$					
s_{1403}		CAL	User	Num_Lock ( )	Main_ Keypad_B		s_{1404}
	r_{1404a}						
		$\propto_{1403}$					
s_{1404}	key_count > 0	CAL	User	Num_Key ( )	Main_ Keypad_B	{key_count = key_count -1;}	s_{1404}
	r_{1404b}						
		$\propto_{1404}$					
s_{1404}		CAL	User	Num_Lock ( )	Main_ Keypad_B		s_{1403}

Figure 14-6. Relation $ASSMR_{KB}$ of the Advanced SBC State Machine $ASSM_{KB}$

14-9 Case Study Two

In this case study, “Free Lottery” has an external actor: “Customer” and two blocks: “Lottery_UI”, “Lottery_Database”. Each customer can draw 3 times for free. We will introduce a draw_count variable, which is initialized to 3, and will be reduced every time the customer draws. The component operation diagram of “Free Lottery” is shown in Figure 14-7.

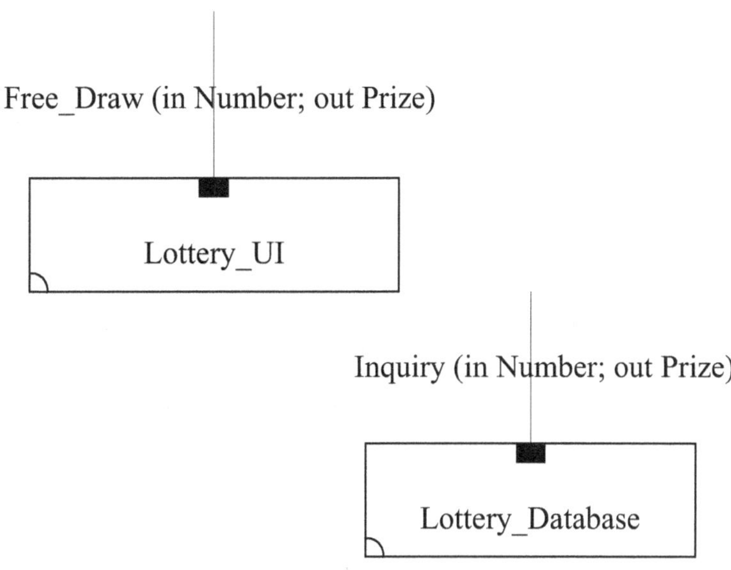

Figure 14-7. COD of "Free Lottery"

We can also list the relationships to describe the component operation diagram. Figure 14-8 shows the relation $COD \subseteq L \times \Gamma$ that represents the COD of “Free Lottery”.

L	Γ
Free_Draw (in Number; out Prize)	Lottery_ UI
Inquiry (in Number; out Prize)	Lottery_ Database

Figure 14-8. COD Relation of "Free Lottery"

In “Free Lottery”, the “Customer” actor requires the “Free_Draw (in Number; out Prize)” operation which is provided by the “Lottery_UI” block; the “Lottery_UI” block requires the “Inquiry (in Number; out Prize)” operation which is provided by the “Lottery_Database” block. Figure 14-9 shows all the operation-based value-passing interactions that occur in “Free Lottery”.

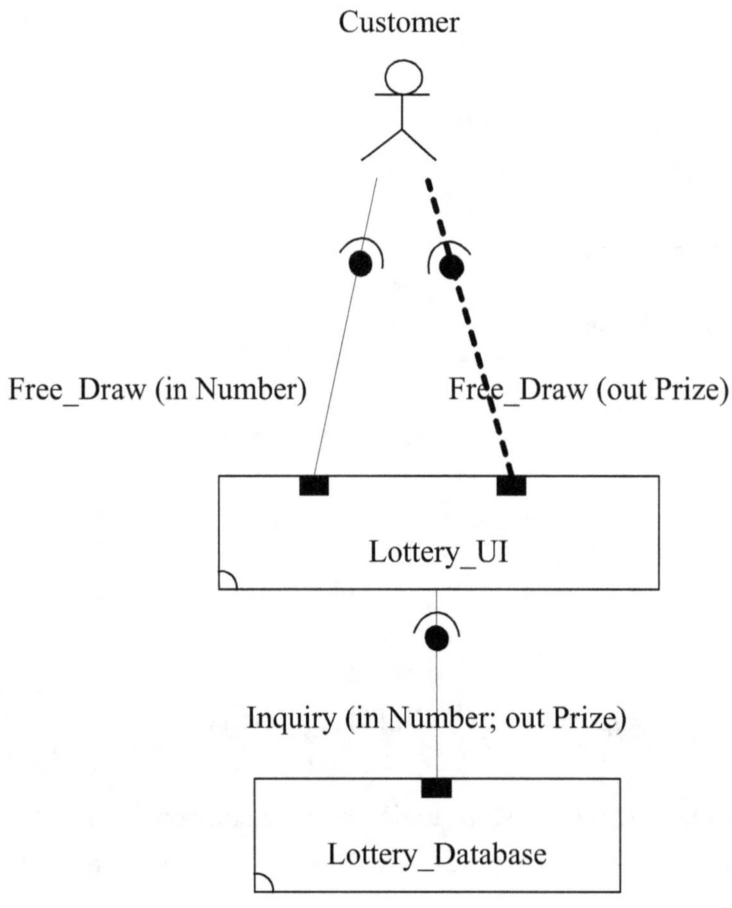

Figure 14-9. Interactions of "Free Lottery"

We can also list the relationships to describe these interactions. Figure 14-10 shows the relation $\Delta \subseteq N \times \Xi \times L \times \Gamma$ that represents the interactions that occur in “Free Lottery”.

Δ			
N	Ξ	L	Γ
a_{1411}			
CAL	Customer	Free_Draw (in Number)	Lottery_ UI
a_{1412}			
CAL	Lottery_ UI	Inquiry (in Number; out Prize)	Lottery_ Database
a_{1413}			
RET	Customer	Free_Draw (out Prize)	Lottery_ UI

Figure 14-10. Interactions Relation of "Free Lottery"

Figure 14-11 describes the advanced SBC state machine "$ASSM_{FL}$" of "Free Lottery".

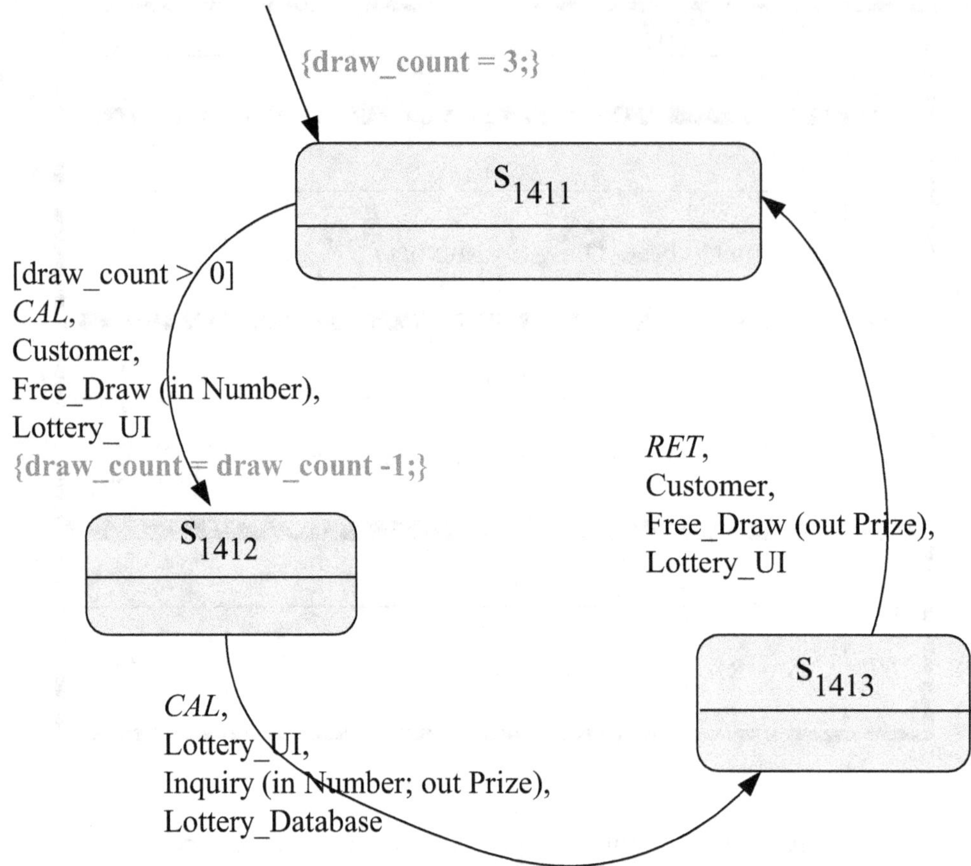

Figure 14-11. Diagram of the Advanced SBC State Machine $ASSM_{FL}$

Code snippets can be attached to the initial transition. Code snippets can also be attached to the prefix of the transitions. The advanced SBC state machine "$ASSM_{FL}$" of "Free Lottery" shows that the draw_count variable is initialized to 3 and decreases by 1 for each free draw.

We can also list the relationships that represent the advanced SBC state machine. Figure 14-12 shows the transition relation "$ASSMR_{FL}$" of the advanced SBC state machine "$ASSM_{FL}$".

Ψ_1	R						Ψ_2
	C	Ω				Π	
		N	Ξ	L	Γ		
	r_{1411}						
s_{1411} {draw_count = 3;}	draw_count > 0	α_{1411}				{draw_count = draw_count -1;}	s_{1412}
		CAL	Customer	Free_Draw (in Number)	Lottery_ UI		
	r_{1412}						
s_{1412}		α_{1412}					s_{1413}
		CAL	Lottery_ UI	Inquiry (in Number; out Prize)	Lottery_ Database		
	r_{1413}						
s_{1413}		α_{1413}					s_{1411}
		RET	Customer	Free_Draw (out Prize)	Lottery_ UI		

Figure 14-12. Relation $ASSMR_{FL}$ of the Advanced SBC State Machine $ASSM_{FL}$

PART III: APPLYING SBC STATE MACHINE TO MODEL-BASED SYSTEMS ENGINEERING

Chapter 15: Projecting the SysML Use Case Diagram from the SBC State Machine

In this chapter, we discuss how to project a SysML use case diagram from the SBC state machine of a system.

15-1 SysML Use Case Diagrams

The simplest SysML use case diagram (UCD) represents the user's interaction with the system, showing the relationship between the user and the different use cases involved. Use case diagrams identify different types of users and different use cases of the system, and are often accompanied by other types of diagrams. The use case is denoted by an oval box containing its name.

The SysML use case diagram of a system UCD_{system} can be defined by an orthogonal composition "$UCD_1 \parallel UCD_2 \parallel \ldots \parallel UCD_m$", as shown in Figure 15-1.

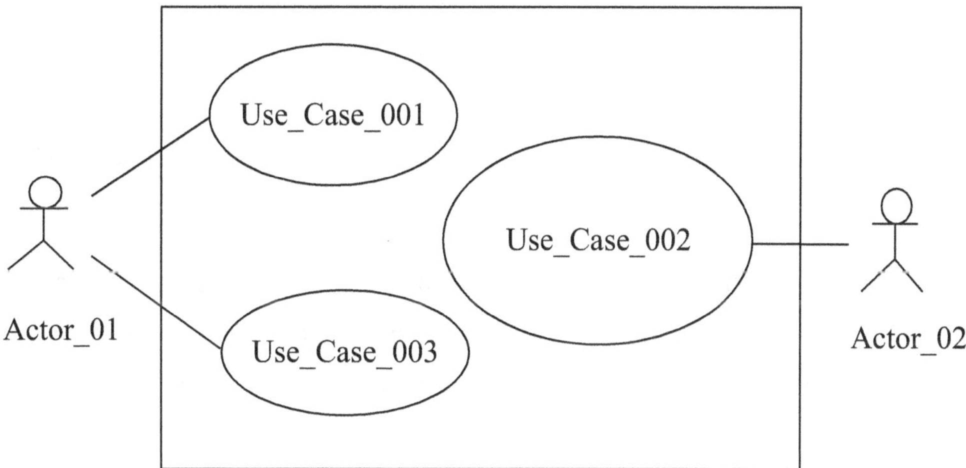

Figure 15-1. SysML Use Case Diagram

15-2 UCD Relation (UCDR) of a System

In SysML, the use case diagram of a system UCD_{system} can be formally defined as "$UCD_1 \| UCD_2 \| \ldots \| UCD_m$" and each UCD_i is represented by a relation $UCDR_i \subseteq B \times U$, where B is a set of "actors" and U is a set of "use cases". Therefore, we get the UCD relation of a system $UCDR_{system} \subseteq B \times U$ defined as "$UCDR_1 \| UCDR_2 \| \ldots \| UCDR_m$" and diagramed as shown in Figure 9-2.

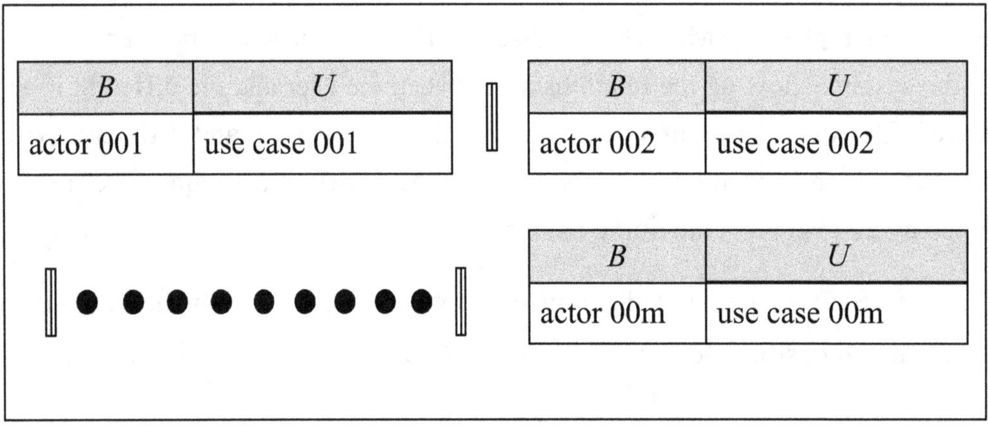

Figure 15-2. Relation $UCDR_{system}$

15-3 Algorithm of Projecting the Use Case Diagram from the SBC State Machine

In SBC state machine, the state expression of a system is represented as SSM_{system} (defined as "$SSM_1 \| SSM_2 \| \ldots \| SSM_m$") with the transition relation $SSMR_{system} \subseteq \Psi_1 \times \Delta \times \Psi_2$ (defined as "$SSMR_1 \| SSMR_2 \| \ldots \| SSMR_m$") as shown in Figure 15-3.

Ψ_1	R	Ψ_2
s_{11}	r_{11}	s_{12}
s_{12}	r_{12}	s_{13}
s_{13}	r_{13}	s_{14}
•	•	•
s_{1n}	r_{1n}	s_{11}

‖

Ψ_1	R	Ψ_2
s_{21}	r_{21}	s_{22}
s_{22}	r_{22}	s_{23}
s_{23}	r_{23}	s_{24}
•	•	•
s_{2n}	r_{2n}	s_{21}

‖ • • • • • • • • ‖

Ψ_1	R	Ψ_2
s_{m1}	r_{m1}	s_{m2}
s_{m2}	r_{m2}	s_{m3}
s_{m3}	r_{m3}	s_{m4}
•	•	•
s_{mn}	r_{mn}	s_{m1}

Figure 15-3. Relation $SSMR_{\text{system}}$

We rewrite the SSM relation of a system as $SSMR_{\text{system}} \subseteq \Psi_1 \times N \times \Xi \times \Lambda \times \Theta \times \Gamma \times \Psi_2$ since the "type 1 or 2 interaction" is defined as a relation $\Delta \subseteq N \times \Xi \times L \times \Gamma$ and the "operation call or operation return signature" is defined as a relation $L \subseteq \Lambda \times \Theta$.

Figure 15-4 shows the algorithm of projecting the UCD relation $UCDR_{system} \subseteq BXU$ from the SSM relation $SSMR_{system} \subseteq \Psi_1 \times N \times \Xi \times \Lambda \times \Theta \times \Gamma \times \Psi_2$.

```
For i = 1, m Loop
  SCANF("%s", @UseCaseName)
  CREATE RELATION UCDR_i (B, U)
  INSERT INTO UCDR_i(B) SELECT Ξ FROM SSMR_i fetch first row only;
  UPDATE UCDR_i SET U = @UseCaseName SELECT * FROM UCDR_i
End Loop;

ORTHOGONALLY COMPOSE All UCDR_i (i.e., ‖_{i=1, m} UCDR_i) to get UCDR_system
```

Figure 15-4. Algorithm of Projecting the UCD Relation from the SSM Relation

Once we have the UCD relation $UCDR_{system}$, it is easy to get a SysML use case diagram of the system.

Chapter 16: Projecting the SysML State Machine from the SBC State Machine

In this chapter, we discuss how to project a SysML state machine from the SBC state machine of a system.

16-1 SysML State Machine

In SysML, the state machine (STM) represents behavior of a system in terms of its transition between states triggered by operation calls. SysML state machine is an object-based variant of Harel statechart, adapted and extended by SysML. The goal of SysML state machine is to overcome the main limitations of traditional finite-state machines while retaining their main benefits. SysML state machine introduce the new concepts of hierarchically nested states and orthogonal regions, while extending the notion of operation calls.

The SysML state machine of a system STM_{system} can be defined by an orthogonal composite state "$STM_1 \parallel STM_2 \parallel \ldots \parallel STM_m$", as shown in Figure 16-1.

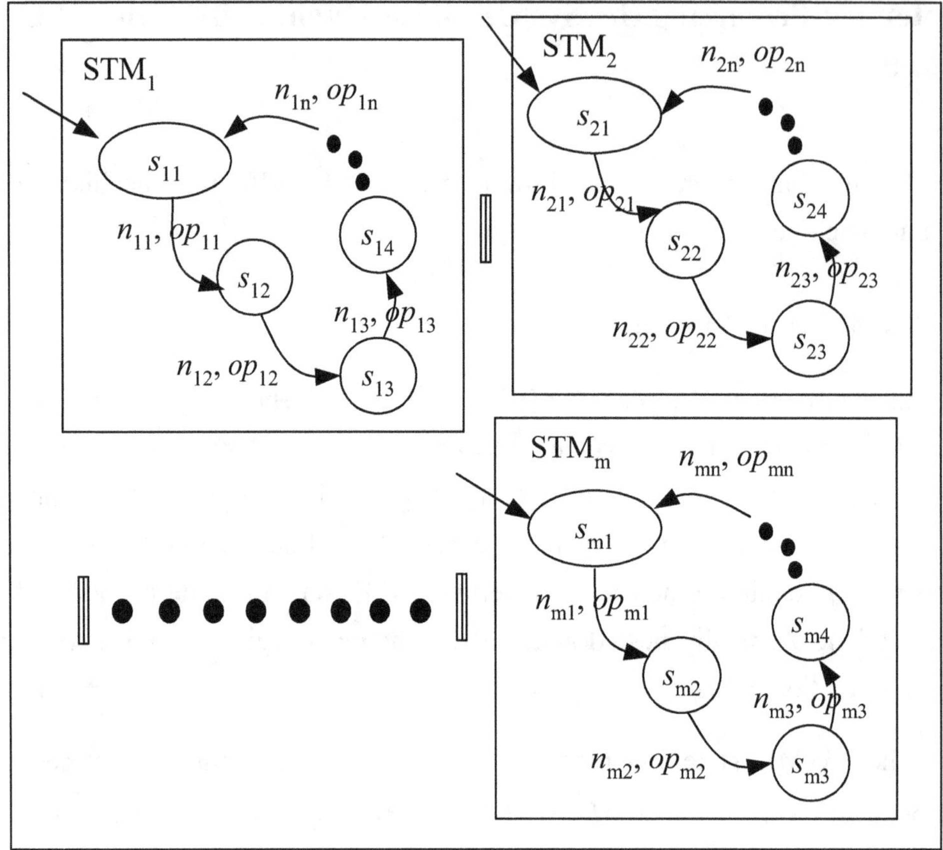

Figure 16-1. SysML State Machine STM_{system}

16-2 STM Relation (STMR) of a System

In SysML, the state machine of a system STM_{system} is formally defined as "$STM_1 \| STM_2 \| \ldots \| STM_m$" and each substate machine STM_i is represented by a relation $STMR_i \subseteq \Psi_1 \times N \times \Lambda \times \Psi_2$, where Ψ is a set of "state expressions" and N is a set of "operation call or operation return tags" and Λ is a set of "operation names" and $(s_{ij}, n, op, s_{ik}) \in STMR_i$ is written as $s_{ij} \xrightarrow{n,\, op} s_{ik}$. Therefore, we get the STM relation of a system $STMR_{\text{system}} \subseteq \Psi_1 \times N \times \Lambda \times \Psi_2$ defined as "$STMR_1 \| STMR_2 \| \ldots \| STMR_m$" and diagramed as shown in Figure 16-2.

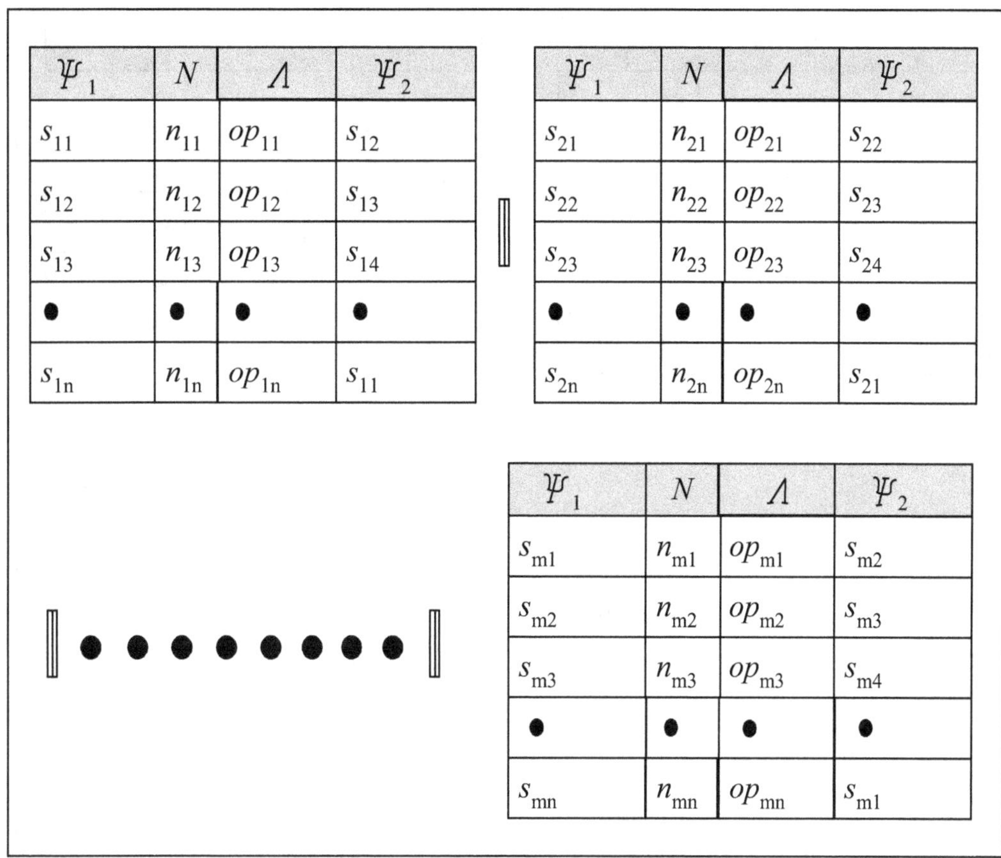

Ψ_1	N	Λ	Ψ_2
s_{11}	n_{11}	op_{11}	s_{12}
s_{12}	n_{12}	op_{12}	s_{13}
s_{13}	n_{13}	op_{13}	s_{14}
•	•	•	•
s_{1n}	n_{1n}	op_{1n}	s_{11}

‖

Ψ_1	N	Λ	Ψ_2
s_{21}	n_{21}	op_{21}	s_{22}
s_{22}	n_{22}	op_{22}	s_{23}
s_{23}	n_{23}	op_{23}	s_{24}
•	•	•	•
s_{2n}	n_{2n}	op_{2n}	s_{21}

‖ • • • • • • • • ‖

Ψ_1	N	Λ	Ψ_2
s_{m1}	n_{m1}	op_{m1}	s_{m2}
s_{m2}	n_{m2}	op_{m2}	s_{m3}
s_{m3}	n_{m3}	op_{m3}	s_{m4}
•	•	•	•
s_{mn}	n_{mn}	op_{mn}	s_{m1}

Figure 16-2. Relation $STMR_{\text{system}}$

16-3 Algorithm of Projecting the SysML State Machine from the SBC State Machine

In O-M-SBC-PA, the state expression of a system is represented by a SBC state machine SSM_{system} (defined as "$SSM_1 \| SSM_2 \| \ldots \| SSM_m$") with the transition relation $SSMR_{\text{system}} \subseteq \Psi_1 \times \Delta \times \Psi_2$ (defined as "$SSMR_1 \| SSMR_2 \| \ldots \| SSMR_m$") as shown in Figure 16-3.

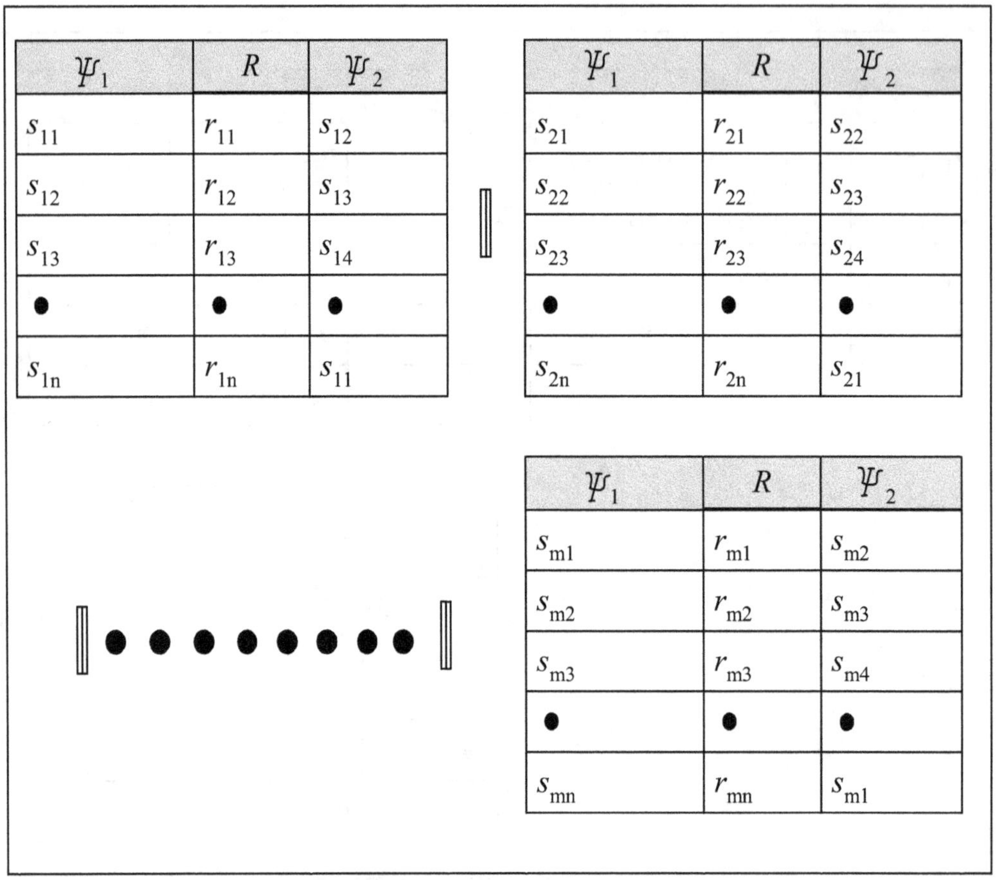

Ψ_1	R	Ψ_2
s_{11}	r_{11}	s_{12}
s_{12}	r_{12}	s_{13}
s_{13}	r_{13}	s_{14}
•	•	•
s_{1n}	r_{1n}	s_{11}

‖

Ψ_1	R	Ψ_2
s_{21}	r_{21}	s_{22}
s_{22}	r_{22}	s_{23}
s_{23}	r_{23}	s_{24}
•	•	•
s_{2n}	r_{2n}	s_{21}

‖ • • • • • • • • ‖

Ψ_1	R	Ψ_2
s_{m1}	r_{m1}	s_{m2}
s_{m2}	r_{m2}	s_{m3}
s_{m3}	r_{m3}	s_{m4}
•	•	•
s_{mn}	r_{mn}	s_{m1}

Figure 16-3. Relation $SSMR_{\text{system}}$

We rewrite the SSM relation of a system as $SSMR_{\text{system}} \subseteq \Psi_1 \times N \times \Xi \times \Lambda \times \Theta \times \Gamma \times \Psi_2$ since the "type 1 or 2 interaction" is defined as a relation $\Delta \subseteq N \times \Xi \times L \times \Gamma$ and the "operation call or operation return signature" is defined as a relation $L \subseteq \Lambda \times \Theta$.

Figure 10-4 shows the algorithm of projecting the STM relation $STMR_{system} \subseteq \Psi_1 \times N \times \Lambda \times \Psi_2$ from the SSM relation $SSMR_{system} \subseteq \Psi_1 \times N \times \Xi \times \Lambda \times \Theta \times \Gamma \times \Psi_2$.

For i = 1, m **Loop**
SELECT Ψ_1, N, Λ, Ψ_2 INTO $STMR_i$ FROM $SSMR_i$;
End Loop;

ORTHOGONALLY COMPOSE All $STMR_i$ (i.e., $\|_{i=1,m} STMR_i$) to get $STMR_{system}$

Figure 16-4. Algorithm of Projecting the STM Relation from the SSM Relation

Once we have the STM relation $STMR_{system}$, it is easy to get a SysML state machine of the system.

Chapter 17: Projecting the SysML Activity Diagram from the SBC State Machine

In this chapter, we discuss how to project a SysML activity diagram from the SBC state machine of a system.

17-1 SysML Activity Diagrams

In SysML, activity diagrams (AD) are intended to model both computational and organizational processes, as well as the data flows intersecting with the related activities. Although activity diagrams primarily show the overall flow of control, they can also include elements showing the flow of data between activities through one or more data stores.

The SysML activity diagram of a system AD_{system} can be defined by an orthogonal composition "$AD_1 \| AD_2 \| \ldots \| AD_m$", as shown in Figure 17-1.

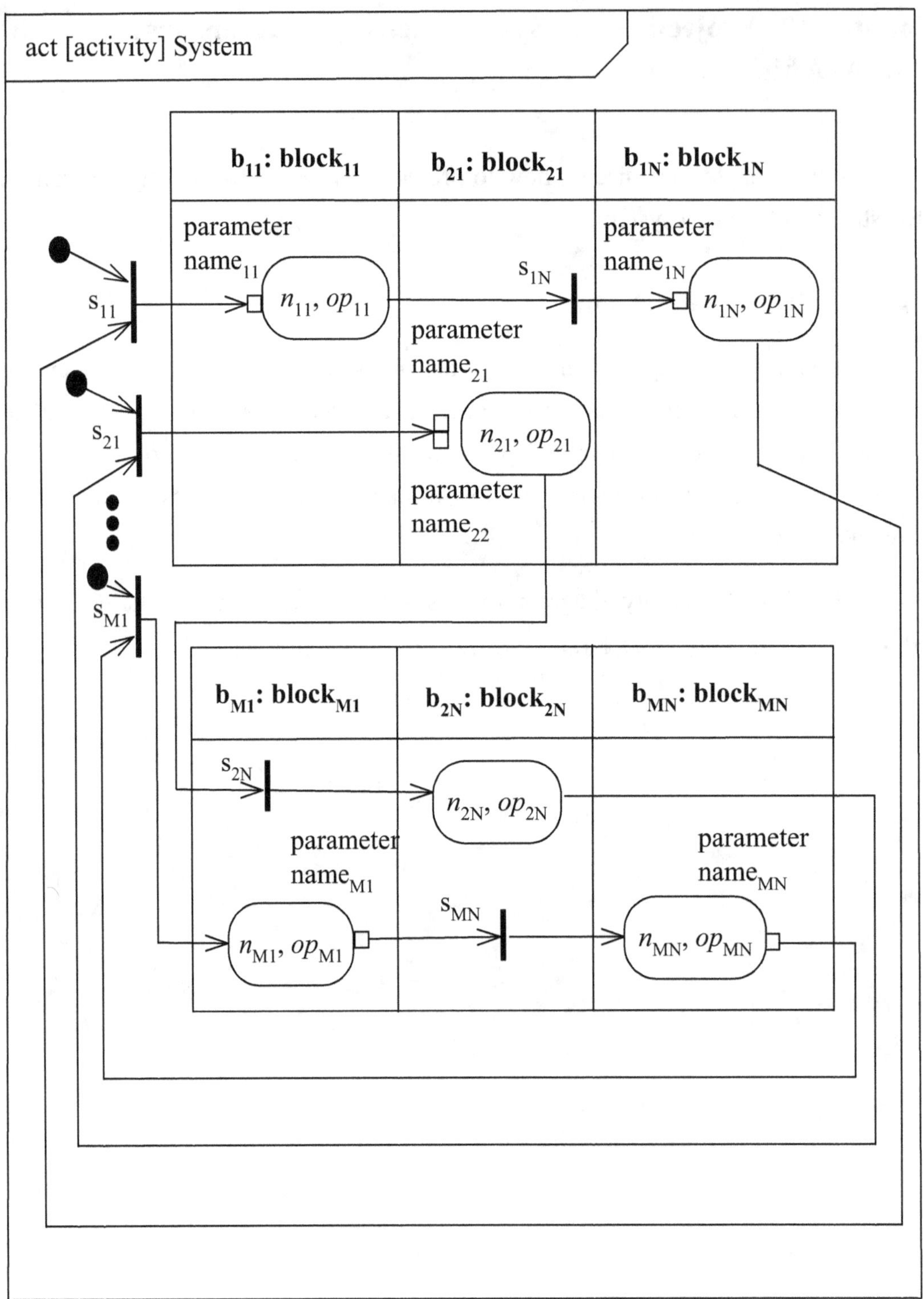

Figure 17-1. SysML Activity Diagram AD_{system}

17-2 AD Relation (ADR) of a System

In SysML, the activity diagram of a system AD_{system} can be formally defined as "$AD_1 \| AD_2 \| \ldots \| AD_m$" and each AD_i is represented by a relation $ADR_i \subseteq \Psi_1 \times N \times \Lambda \times \Theta \times \Gamma \times \Psi_2$, where Ψ is a set of "state expressions" and N is a set of "operation call or operation return tags" and Λ is a set of "operation names" and Θ is a set of "parameter lists" and Γ is a set of "blocks" and $(s_{ij}, n, op, p, b, s_{ik}) \in ADR_i$ is written as $s_{ij} \xrightarrow{n,\, op,\, p,\, b} s_{ik}$. Therefore, we get the AD relation of a system $ADR_{system} \subseteq \Psi_1 \times N \times \Lambda \times \Theta \times \Gamma \times \Psi_2$ defined as "$ADR_1 \| ADR_2 \| \ldots \| ADR_m$" and diagramed as shown in Figure 17-2.

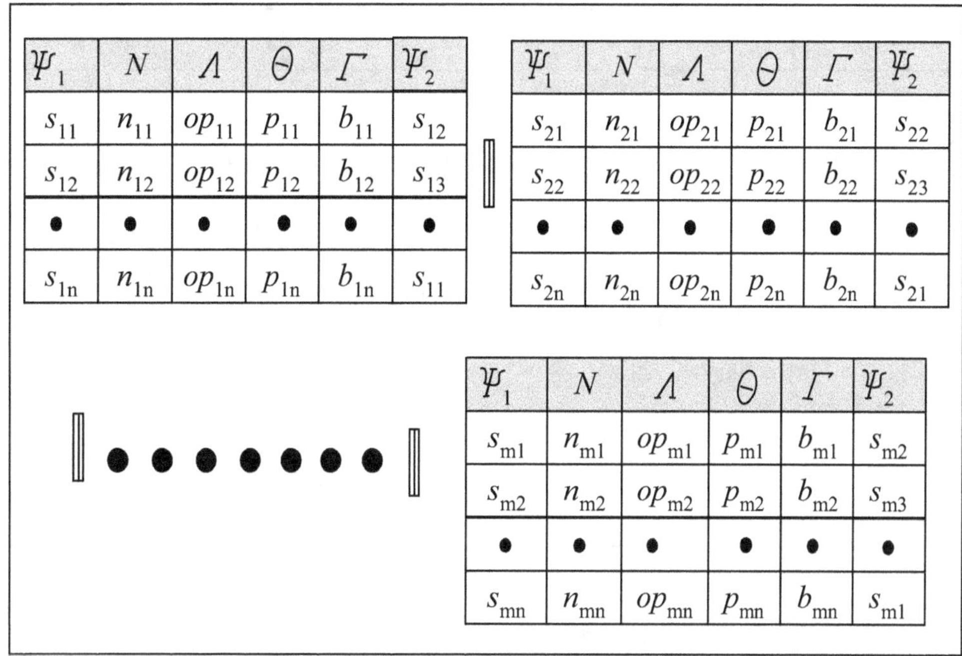

Ψ_1	N	Λ	Θ	Γ	Ψ_2
s_{11}	n_{11}	op_{11}	p_{11}	b_{11}	s_{12}
s_{12}	n_{12}	op_{12}	p_{12}	b_{12}	s_{13}
•	•	•	•	•	•
s_{1n}	n_{1n}	op_{1n}	p_{1n}	b_{1n}	s_{11}

$\|$

Ψ_1	N	Λ	Θ	Γ	Ψ_2
s_{21}	n_{21}	op_{21}	p_{21}	b_{21}	s_{22}
s_{22}	n_{22}	op_{22}	p_{22}	b_{22}	s_{23}
•	•	•	•	•	•
s_{2n}	n_{2n}	op_{2n}	p_{2n}	b_{2n}	s_{21}

$\|$ • • • • • • • $\|$

Ψ_1	N	Λ	Θ	Γ	Ψ_2
s_{m1}	n_{m1}	op_{m1}	p_{m1}	b_{m1}	s_{m2}
s_{m2}	n_{m2}	op_{m2}	p_{m2}	b_{m2}	s_{m3}
•	•	•	•	•	•
s_{mn}	n_{mn}	op_{mn}	p_{mn}	b_{mn}	s_{m1}

Figure 17-2. Relation ADR_{system}

17-3 Algorithm of Projecting the Activity Diagram from the SBC State Machine

In O-M-SBC-PA, the state expression of a system is represented by a SBC state machine SSM_{system} (defined as "$SSM_1 \| SSM_2 \| \ldots \| SSM_m$") with the transition relation $SSMR_{system} \subseteq \Psi_1 \times \Delta \times \Psi_2$ (defined as "$SSMR_1 \| SSMR_2 \| \ldots \| SSMR_m$") as shown in Figure 17-3.

Ψ_1	R	Ψ_2
s_{11}	r_{11}	s_{12}
s_{12}	r_{12}	s_{13}
s_{13}	r_{13}	s_{14}
•	•	•
s_{1n}	r_{1n}	s_{11}

‖

Ψ_1	R	Ψ_2
s_{21}	r_{21}	s_{22}
s_{22}	r_{22}	s_{23}
s_{23}	r_{23}	s_{24}
•	•	•
s_{2n}	r_{2n}	s_{21}

‖ • • • • • • • • ‖

Ψ_1	R	Ψ_2
s_{m1}	r_{m1}	s_{m2}
s_{m2}	r_{m2}	s_{m3}
s_{m3}	r_{m3}	s_{m4}
•	•	•
s_{mn}	r_{mn}	s_{m1}

Figure 17-3. Relation $SSMR_{\text{system}}$

We rewrite the SSM relation of a system as $SSMR_{\text{system}} \subseteq \Psi_1 \times N \times \Xi \times \Lambda \times \Theta \times \Gamma \times \Psi_2$ since the "type 1 or 2 interaction" is defined as a relation $\Delta \subseteq N \times \Xi \times L \times \Gamma$ and the "operation call or operation return signature" is defined as a relation $L \subseteq \Lambda \times \Theta$.

Figure 17-4 shows the algorithm of projecting the AD relation $ADR_{system} \subseteq \Psi_1 \times N \times \Lambda \times \Theta \times \Gamma \times \Psi_2$ from the SSM relation $SSMR_{system} \subseteq \Psi_1 \times N \times \Xi \times \Lambda \times \Theta \times \Gamma \times \Psi_2$.

For i = 1, m **Loop**
SELECT Ψ_1, N, Λ, Θ, Γ, Ψ_2 INTO ADR_i FROM $SSMR_i$;
End Loop;

ORTHOGONALLY COMPOSE All ADR_i (i.e., $\|_{i=1,m} ADR_i$) to get ADR_{system}

Figure 17-4. Algorithm of Projecting the AD Relation from the SSM Relation

Once we have the AD relation ADR_{system}, it is easy to get a SysML activity diagram of the system.

Chapter 18: Projecting the SysML Sequence Diagram from the SBC State Machine

In this chapter, we discuss how to project a SysML sequence diagram from the SBC state machine of a system.

18-1 SysML Sequence Diagrams

SysML sequence diagrams (SqD) are interaction diagrams that detail how to perform operations. They capture interactions between blocks in a collaborative environment. Sequence diagrams are time focuses that visually display the order of interactions by using the vertical axis of the diagram to indicate when and when the message was sent. Sequence diagrams are sometimes called event diagrams or event scenarios.

The SysML sequence diagram of a system SqD_{system} can be defined by an orthogonal composition "$SqD_1 \| SqD_2 \| \ldots \| SqD_{\text{m}}$", as shown in Figure 18-1.

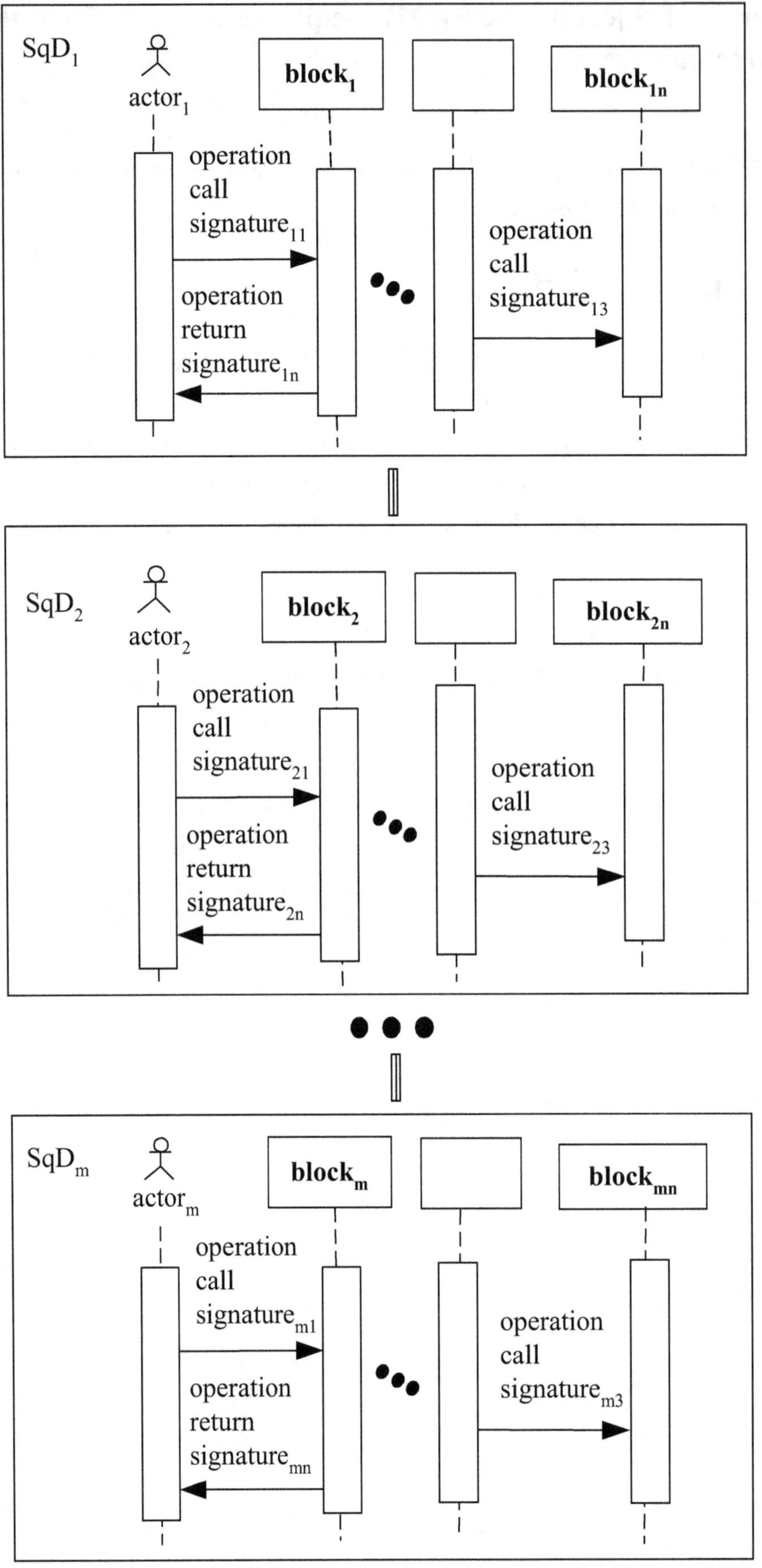

Figure 18-1. SysML Sequence Diagram SqD_{system}

18-2 SqD Relation (SqDR) of a System

In SysML, the sequence diagram of a system SqD_{system} is formally defined as "$SqD_1 \parallel SqD_2 \parallel \ldots \parallel SqD_m$" and each SqD_i is represented by a relation $SqDR_i \subseteq E \times N \times \Xi \times \Lambda \times \Theta \times \Gamma$, where E is a set of "execution orders" and N is a set of "operation call or operation return tags" and Ξ is a set of "actors or blocks" and Λ is a set of "operation names" and Θ is a set of "parameter lists" and Γ is a set of "blocks". Therefore, we get the SqD relation of a system $SqDR_{system} \subseteq E \times N \times \Xi \times \Lambda \times \Theta \times \Gamma$ defined as "$SqDR_1 \parallel SqDR_2 \parallel \ldots \parallel SqDR_m$" and diagramed as shown in Figure 18-2.

E	N	Ξ	Λ	Θ	Γ
1	n_{11}	ρ_{11}	op_{11}	p_{11}	b_{11}
2	n_{12}	ρ_{12}	op_{12}	p_{12}	b_{12}
•	•	•	•	•	•
n	n_{1n}	ρ_{1n}	op_{1n}	p_{1n}	b_{1n}

E	N	Ξ	Λ	Θ	Γ
1	n_{21}	ρ_{21}	op_{21}	p_{21}	b_{21}
2	n_{22}	ρ_{22}	op_{22}	p_{22}	b_{22}
•	•	•	•	•	•
n	n_{2n}	ρ_{2n}	op_{2n}	p_{2n}	b_{2n}

E	N	Ξ	Λ	Θ	Γ
1	n_{m1}	ρ_{m1}	op_{m1}	p_{m1}	b_{m1}
2	n_{m2}	ρ_{m2}	op_{m2}	p_{m2}	b_{m2}
•	•	•	•	•	•
n	n_{mn}	ρ_{mn}	op_{mn}	p_{mn}	b_{mn}

Figure 18-2. Relation $SqDR_{\text{system}}$

18-3 Algorithm of Projecting the Sequence Diagram from the SBC State Machine

In O-M-SBC-PA, the state expression of a system is represented by a SBC state machine SSM_{system} (defined as "$SSM_1 \| SSM_2 \| \ldots \| SSM_m$") with the transition relation $SSMR_{system} \subseteq \Psi_1 \times \Delta \times \Psi_2$ (defined as "$SSMR_1 \| SSMR_2 \| \ldots \| SSMR_m$") as shown in Figure 18-3.

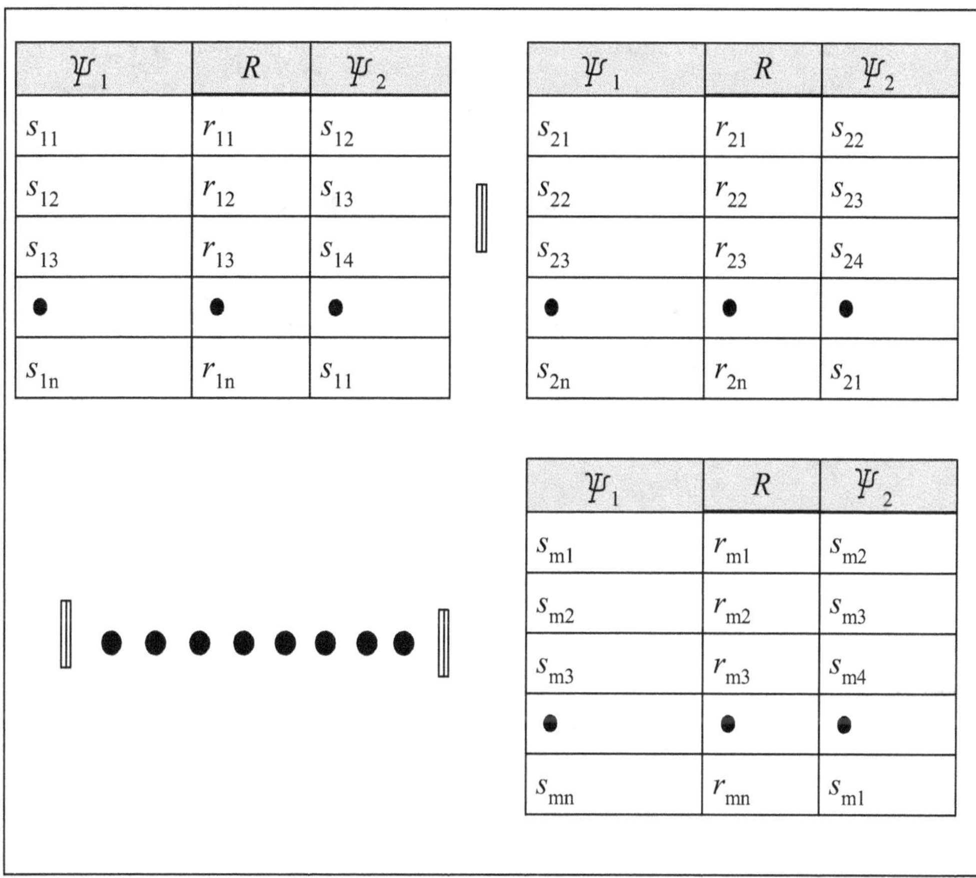

Ψ_1	R	Ψ_2
s_{11}	r_{11}	s_{12}
s_{12}	r_{12}	s_{13}
s_{13}	r_{13}	s_{14}
•	•	•
s_{1n}	r_{1n}	s_{11}

$\|$

Ψ_1	R	Ψ_2
s_{21}	r_{21}	s_{22}
s_{22}	r_{22}	s_{23}
s_{23}	r_{23}	s_{24}
•	•	•
s_{2n}	r_{2n}	s_{21}

$\|$ • • • • • • • • $\|$

Ψ_1	R	Ψ_2
s_{m1}	r_{m1}	s_{m2}
s_{m2}	r_{m2}	s_{m3}
s_{m3}	r_{m3}	s_{m4}
•	•	•
s_{mn}	r_{mn}	s_{m1}

Figure 18-3. Relation $SSMR_{system}$

We rewrite the SSM relation of a system as $SSMR_{system} \subseteq \Psi_1 \times N \times \Xi \times \Lambda \times \Theta \times \Gamma \times \Psi_2$ since the "type 1 or 2 interaction" is defined as a relation $\Delta \subseteq N \times \Xi \times L \times \Gamma$ and the "operation call or operation return signature" is defined as a relation $L \subseteq \Lambda \times \Theta$.

Figure 18-4 shows the algorithm of projecting the SqD relation $SqDR_{system} \subseteq E \times N \times \Xi \times \Lambda \times \Theta \times \Gamma$ from the SSM relation $SSMR_{system} \subseteq \Psi_1 \times N \times \Xi \times \Lambda \times \Theta \times \Gamma \times \Psi_2$.

For i = 1, m **Loop**
CREATE RELATION $SqDR_i$ (E int IDENTITY(1,1), $N, \Xi, \Lambda, \Theta, \Gamma$);
INSERT INTO $SqDR_i$ ($N, \Xi, \Lambda, \Theta, \Gamma$) SELECT $N, \Xi, \Lambda, \Theta, \Gamma$ FROM $SSMR_i$;
End Loop;

ORTHOGONALLY COMPOSE All $SqDR_i$ (i.e., $\Vert_{i=1,m}\ SqDR_i$) to get $SqDR_{system}$

Figure 18-4. Algorithm of Projecting the SqD Relation from the SSM Relation

Once we have the SqD relation $SqDR_{system}$, it is easy to get the SysML sequence diagram of the system.

Chapter 19: Projecting the SysML Internal Block Diagram from the SBC State Machine

In this chapter, we discuss how to project a SysML internal block diagram from the SBC state machine of a system.

19-1 SysML Internal Block Diagrams

An internal block diagram (IBD) in Systems Modeling Language (SysML) is a static structure diagram that describes the connection between parts of a block, as shown in Figure 19-1.

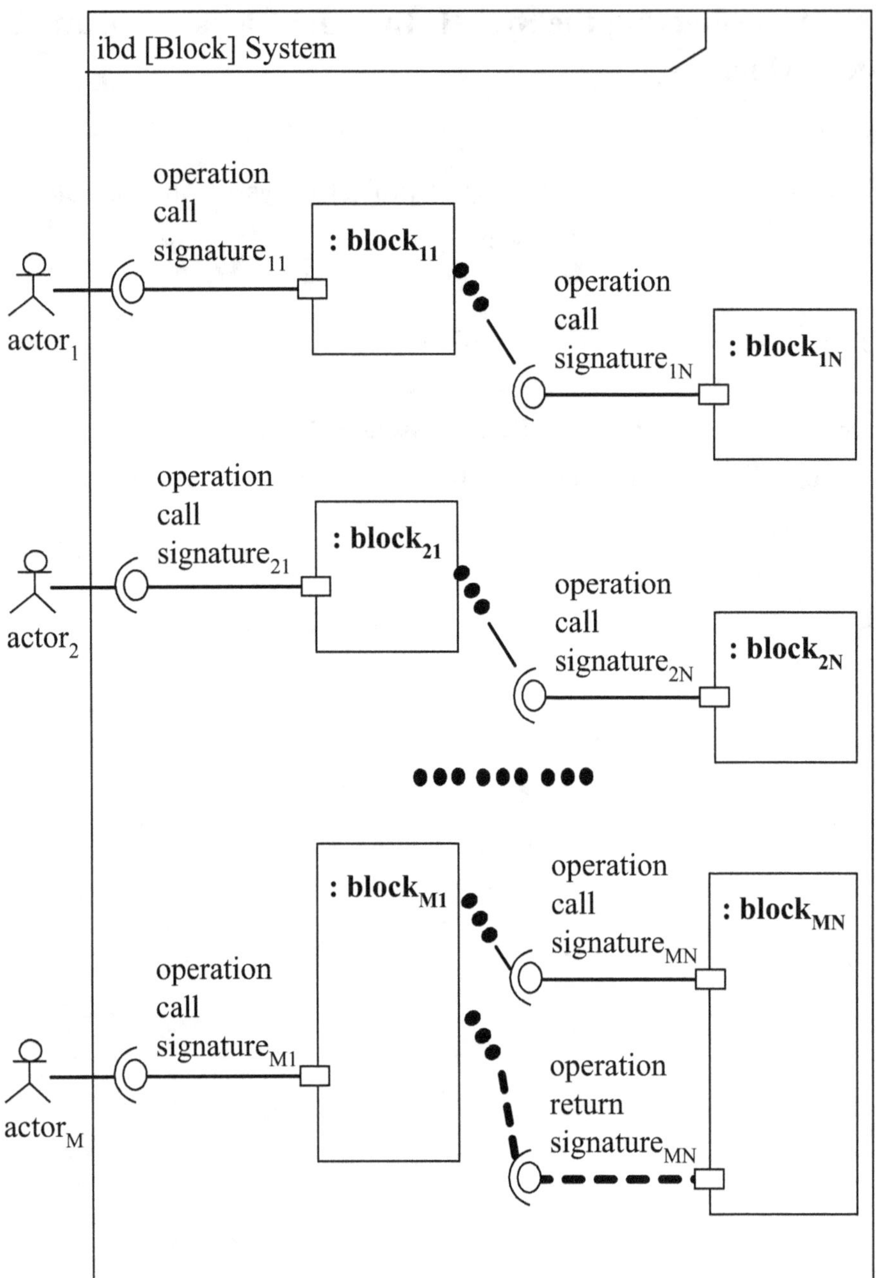

Figure 19-1. SysML Internal Block Diagram IBD_{system}

19-2 IBD Relation (IBDR) of a System

In SysML, the internal block diagram of a system IBD_{system} is formally represented by a relation $IBDR_{\text{system}} \subseteq N \times \Xi \times \Lambda \times \Theta \times \Gamma$, where N is a set of "operation call or operation return tags" and Ξ is a set of "external environment's actors or blocks" and Λ is a set of "operation names" and Θ is a set of "parameter lists" and Γ is a set of "blocks", as shown in Figure 19-2.

N	Ξ	Λ	Θ	Γ
n_{11}	ρ_{11}	op_{11}	p_{11}	b_{11}
n_{12}	ρ_{12}	op_{12}	p_{12}	b_{12}
•	•	•	•	•
n_{1n}	ρ_{1n}	op_{1n}	p_{1n}	b_{1n}

N	Ξ	Λ	Θ	Γ
n_{21}	ρ_{21}	op_{21}	p_{21}	b_{21}
n_{22}	ρ_{22}	op_{22}	p_{22}	b_{22}
•	•	•	•	•
n_{2n}	ρ_{2n}	op_{2n}	p_{2n}	b_{2n}

• • •

N	Ξ	Λ	Θ	Γ
n_{m1}	ρ_{m1}	op_{m1}	p_{m1}	b_{m1}
n_{m2}	ρ_{m2}	op_{m2}	p_{m2}	b_{m2}
•	•	•	•	•
n_{mn}	ρ_{mn}	op_{mn}	p_{mn}	b_{mn}

Figure 19-2. Relation $IBDR_{system}$

19-3 Algorithm of Projecting the Internal Block Diagram from the SBC State Machine

In O-M-SBC-PA, the state expression of a system is represented by a SBC state machine SSM_{system} (defined as "$SSM_1 \parallel SSM_2 \parallel \ldots \parallel SSM_m$") with the transition relation $SSMR_{system} \subseteq \Psi_1 \times \Delta \times \Psi_2$ (defined as "$SSMR_1 \| SSMR_2 \| \ldots \| SSMR_m$") as shown in Figure 19-3.

Ψ_1	R	Ψ_2
s_{11}	r_{11}	s_{12}
s_{12}	r_{12}	s_{13}
s_{13}	r_{13}	s_{14}
•	•	•
s_{1n}	r_{1n}	s_{11}

$\|$

Ψ_1	R	Ψ_2
s_{21}	r_{21}	s_{22}
s_{22}	r_{22}	s_{23}
s_{23}	r_{23}	s_{24}
•	•	•
s_{2n}	r_{2n}	s_{21}

$\|$ • • • • • • • • $\|$

Ψ_1	R	Ψ_2
s_{m1}	r_{m1}	s_{m2}
s_{m2}	r_{m2}	s_{m3}
s_{m3}	r_{m3}	s_{m4}
•	•	•
s_{mn}	r_{mn}	s_{m1}

Figure 19-3. Relation $SSMR_{system}$

We rewrite the SSM relation of a system as $SSMR_{\text{system}} \subseteq \Psi_1 \times N \times \Xi \times \Lambda \times \Theta \times \Gamma \times \Psi_2$ since the "type 1 or 2 interaction" is defined as a relation $\Delta \subseteq N \times \Xi \times L \times \Gamma$ and the "operation call or operation return signature" is defined as a relation $L \subseteq \Lambda \times \Theta$.

Figure 19-4 shows the algorithm of projecting the IBD relation $IBDR_{\text{system}} \subseteq N \times \Xi \times \Lambda \times \Theta \times \Gamma$ from the SSM relation $SSMR_{\text{system}} \subseteq \Psi_1 \times N \times \Xi \times \Lambda \times \Theta \times \Gamma \times \Psi_2$.

```
For i = 1, m Loop
  SELECT N, Ξ, Λ, Θ, Γ INTO IBDR_i (N, Ξ, Λ, Θ, Γ) FROM SSMRi;
  INSERT INTO IBDR_1~m (N, Ξ, Λ, Θ, Γ) SELECT * FROM IBDR_i;
End Loop;

SELECT DISTINCT * INTO IBDR_system FROM IBDR_1~m
```

Figure 19-4. Algorithm of Projecting the IBD Relation from the SSM Relation

Once we have the IBD relation $IBDR_{\text{system}}$, it is easy to get a SysML internal block diagram of the system.

PART IV: CASE STUDY

Chapter 20: Vending Machine

A vending machine is an automated machine that sells products to customers. A customer inserts enough money for a product and selects a specific product; then, the vending machine dispenses the selected product and dispenses change if necessary. A vending machine is a rather complex system with several features. Customers interact with the coin receptacle and buttons of the vending machine and the vendor needs to refill the vending products and change. To create a system modeling of a vending machine, we begin by describing the requirements of the vending machine.

Behaviors of the vending machine consist of: a) behavior of *Getting_Customer_Payment*, b) behavior of *Returning_Customer_Payment*, c) behavior of *Delivering_Customer_Selection*, d) behavior of *Refilling_Product_Store* and e) behavior of *Refilling_Coin_Store.*

In the behavior of *Getting_Customer_Payment*, a customer shall use the *Coin_Receptacle* block to deposit a nickel, dime, or quarter coin. The *Vending Machine* will only initiate payment computation or product selection process after a valid coin is detected. In the behavior of *Returning_Customer_Payment*, a customer shall use the *Return_Payment_Button* block to get his deposit payment returned. In the behavior of *Delivering_Customer_Selection*, a customer shall use the *Product_Selection_Buttons* block to make his product selection. After that, the *Vending Machine* will dispense the selected product to the customer. In the behavior of *Refilling_Product_Store*, a vendor will refill some vending products into the *Product_Store* block. In the behavior of *Refilling_Coin_Store*, a vendor will refill some change coins into the *Coin_Store* block.

20-1 State Expression of the Vending Machine

In SBC state machine, the state expression of the vending machine, s_{VM}, is defined as "$\mathbf{fix}(X_1=g_{11}\bullet v_{12}\bullet v_{13}\bullet v_{14}\bullet X_1)\parallel\mathbf{fix}(X_2=g_{21}\bullet v_{22}\bullet v_{23}\bullet v_{24}\bullet g_{25}\bullet v_{14}\bullet X_2)\parallel\mathbf{fix}(X_3=g_{31}\bullet v_{32}\bullet v_{33}\bullet v_{34}\bullet g_{35}\bullet v_{23}\bullet v_{24}\bullet g_{25}\bullet v_{14}\bullet X_3)\parallel\mathbf{fix}(X_4=g_{41}\bullet X_4)\parallel\mathbf{fix}(X_5=g_{51}\bullet X_5)$".

20-2 COD of the Vending Machine

The component operation diagram (COD) of the vending machine is shown in Figure 20-1.

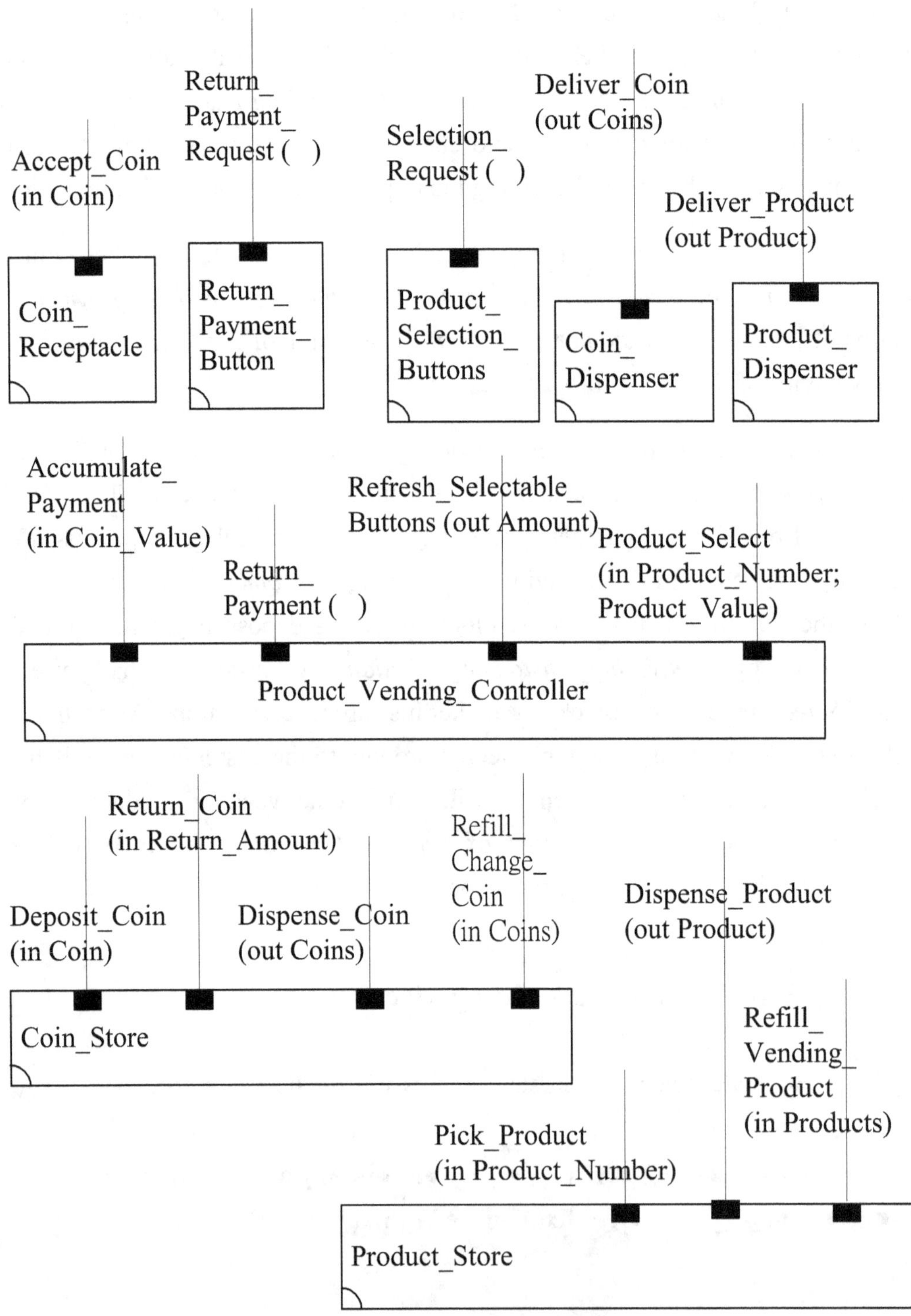

Figure 20-1. COD of the Vending Machine

We can also list the relationships to describe the component operation diagram. Figure 20-2 shows the relation $COD \subseteq \Lambda \times \Theta \times \Gamma$ that represents the COD of the vending machine.

Λ	Θ	Γ
Accept_ Coin	in Coin	:Coin_ Receptacle
Deposit_ Coin	in Coin	:Coin_ Store
Accumulate_ Payment	in Coin_Value	:Product_ Vending_ Controller
Refresh_ Selectable_ Buttons	out Amount	:Product_ Vending_ Controller
Return_ Payment_ Request		:Return_ Payment_ Button
Return_ Payment		:Product_ Vending_ Controller
Return_ Coin	in Return_Amount	:Coin_ Store
Dispense_ Coin	out Coins	:Coin_ Store
Deliver_ Coin	out Coins	:Coin_ Dispenser

Figure 20-2. COD Relation of "Vending Machine"

Λ	Θ	Γ
Selection_ Request		:Product_ Vending_ Controller
Product_ Select	in Product_Number ; Product_Value	:Product_ Vending_ Controller
Pick_ Product	in Product_Number	:Product_ Store
Dispense_ Product	out Product	:Product_ Store
Deliver_ Product	out Product	:Product_ Dispenser
Refill_ Vending_ Product	in Products	
Refill_ Change_ Coin	in Coins	

Figure 20-2 (continued). COD Relation of "Vending Machine"

20-3 Interactions of the Vending Machine

The vending machine has two external actors: "Customer", "Vendor" and eight blocks: "Coin_Receptacle", "Return_Payment_Button", "Product_Selection_Buttons", "Coin_Dispenser", "Product_Dispenser", "Product_Vending_Controller", "Coin_Store", "Product_ Store". Figure 20-3 shows those interactions that occur in the vending machine.

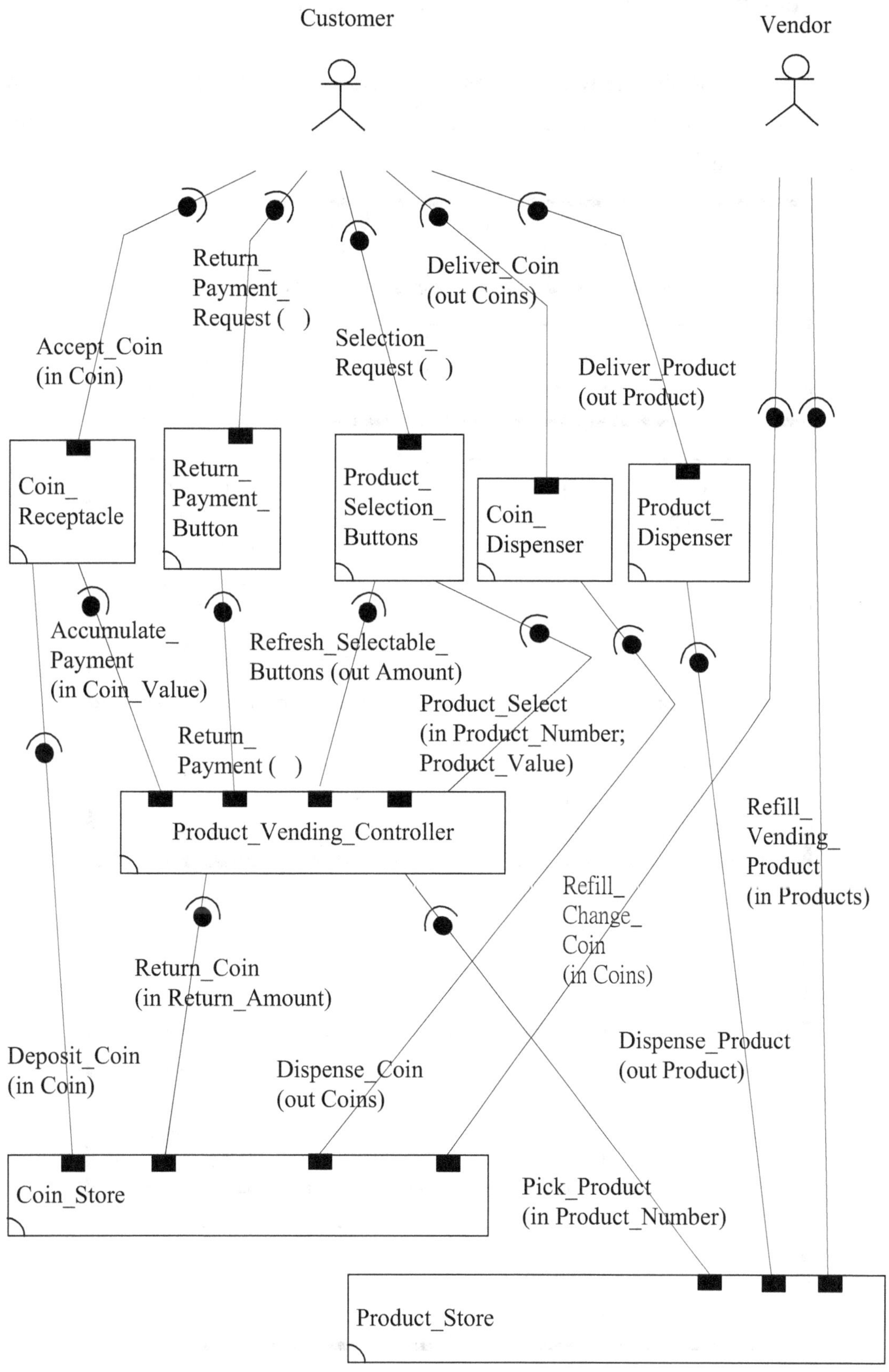

Figure 20-3. Interactions of the Vending Machine

20-4 Interactions Relation of the Vending Machine

We can also list the relationships to describe these interactions. Figure 20-4 shows the relation $\Delta \subseteq N \times \Xi \times \Lambda \times \Theta \times \Gamma$ that represents the interactions that occur in the vending machine.

Δ				
N	Ξ	Λ	Θ	Γ
		g_{11}		
CAL	Customer	Accept_ Coin	in Coin	:Coin_ Receptacle
		v_{12}		
CAL	:Coin_ Receptacle	Deposit_ Coin	in Coin	:Coin_ Store
		v_{13}		
CAL	:Coin_ Receptacle	Accumulate_ Payment	in Coin_Value	:Product_ Vending_ Controller
		v_{14}		
CAL	:Product_ Selection_ Buttons	Refresh_ Selectable_ Buttons	out Amount	:Product_ Vending_ Controller
		g_{21}		
CAL	Customer	Return_ Payment_ Request		:Return_ Payment_ Button
		v_{22}		
CAL	:Return_ Payment_ Button	Return_ Payment		:Product_ Vending_ Controller
		v_{23}		
CAL	:Product_ Vending_ Controller	Return_ Coin	in Return_Amount	:Coin_ Store

Figure 20-4. Interactions Relation of the Vending Machine

Δ				
N	Ξ	Λ	Θ	Γ
		v_{24}		
CAL	:Coin_ Dispenser	Dispense_ Coin	out Coins	:Coin_ Store
		g_{25}		
CAL	Customer	Deliver_ Coin	out Coins	:Coin_ Dispenser
		g_{31}		
CAL	Customer	Selection_ Request		:Product_ Selection_ Buttons
		v_{32}		
CAL	:Product Selection Buttons	Product_ Select	in Product_Number; Product_Value	:Product Vending Controller
		v_{33}		
CAL	:Product Vending Controller	Pick_ Product	in Product_Number	:Product_ Store
		v_{34}		
CAL	:Product_ Dispenser	Dispense_ Product	out Product	:Product_ Store
		g_{35}		
CAL	Customer	Deliver_ Product	out Product	:Product_ Dispenser
		g_{41}		
CAL	Vendor	Refill_ Vending_ Product	in Products	:Product_ Store
		g_{51}		
CAL	Vendor	Refill_ Change_ Coin	in Coins	:Coin_ Store

Figure 20-4 (continued). Interactions Relation of the Vending Machine

20-5 SBC State Machine of the Vending Machine

We use the SBC state machine SSM_{VM} defined as "$SSM_1 \| SSM_2 \| SSM_3 \| SSM_4 \| SSM_5$" to represent the execution of the state expression of the vending machine, as shown in Figure 20-5.

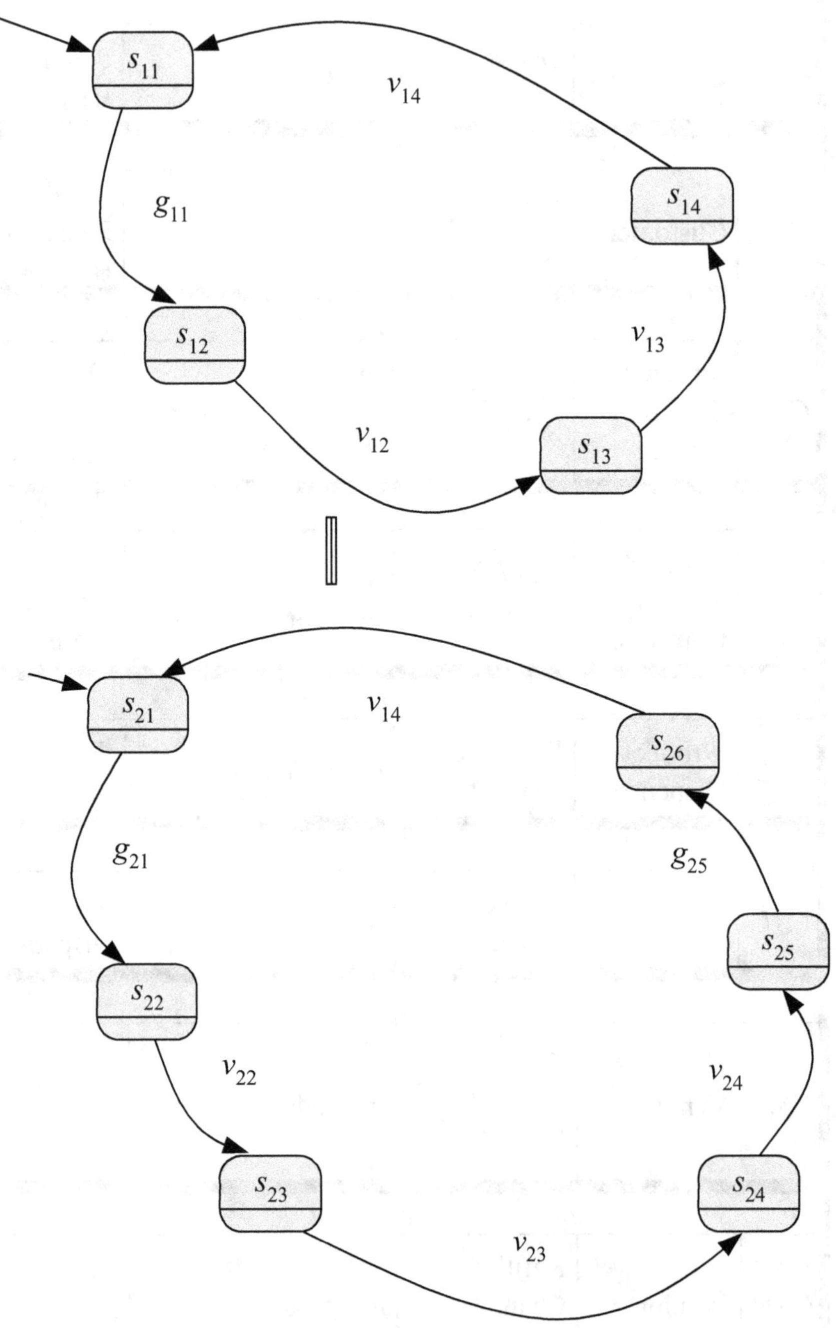

Figure 20-5. SBC State Machine SSM_{VM}

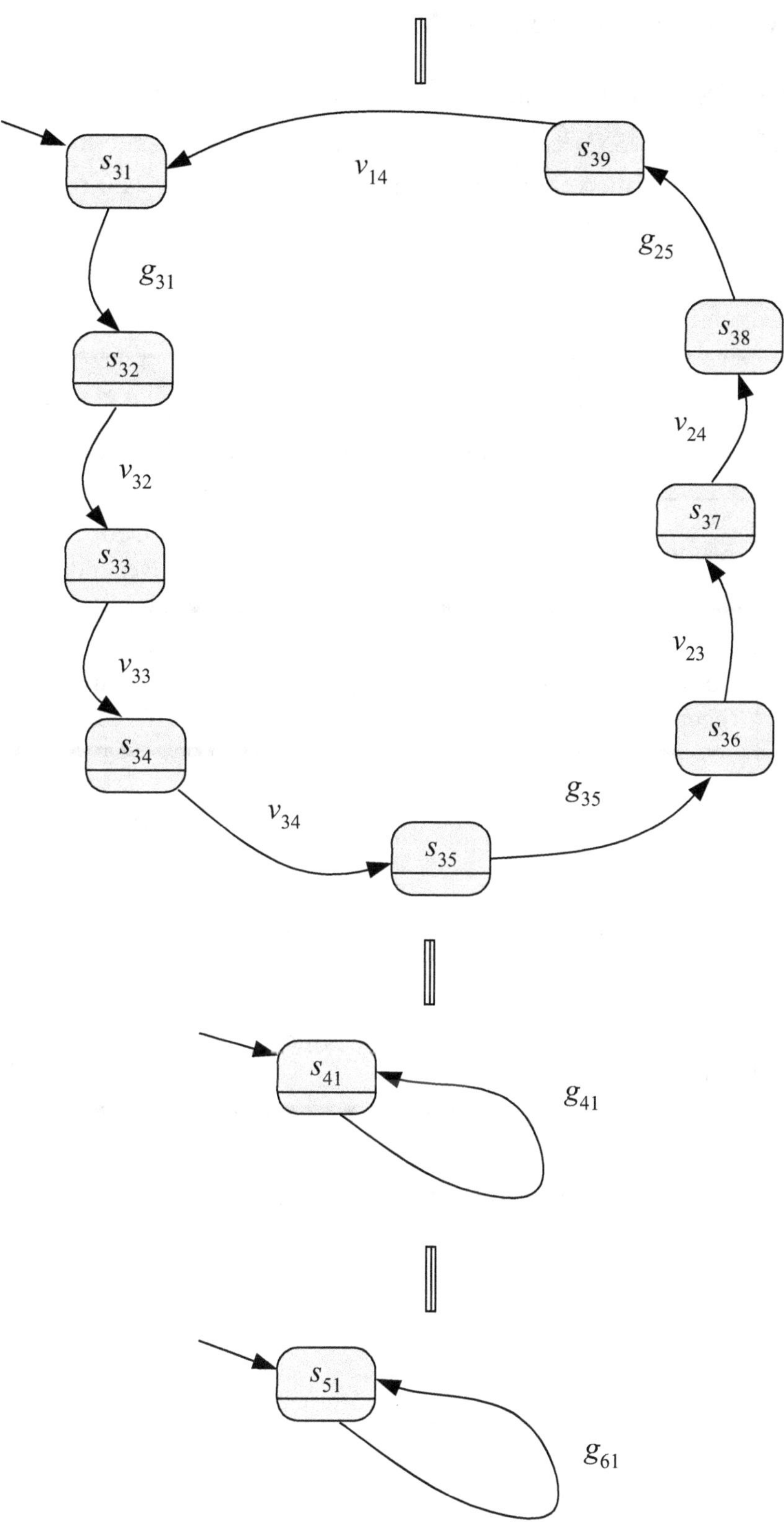

Figure 20-5 (continued). SBC State Machine SSM_{VM}

20-6 SSM Relation of the Vending Machine

We use a SSM relation $SSMR_{VM} \subseteq \Psi_1 \text{ X } N \text{ X } \Xi \text{ X } \Lambda \text{ X } \Theta \text{ X } \Gamma \text{ X } \Psi_2$ defined as "$SSMR_1 \| SSMR_2 \| SSMR_3 \| SSMR_4 \| SSMR_5$" to represent the SBC state machine SSM_{VM} of the state expression of the vending machine, as shown in Figure 20-6.

Ψ_1	Δ					Ψ_2
	N	Ξ	Λ	Θ	Γ	
	g_{11}					
s_{11}	*CAL*	Customer	Accept_ Coin	in Coin	:Coin_ Receptacle	s_{12}
	v_{12}					
s_{12}	*CAL*	:Coin_ Receptacle	Deposit_ Coin	in Coin	:Coin_ Store	s_{13}
	v_{13}					
s_{13}	*CAL*	:Coin_ Receptacle	Accumulate_ Payment	in Coin_Value	:Product_ Vending_ Controller	s_{14}
	v_{14}					
s_{14}	*CAL*	:Product_ Selection_ Buttons	Refresh_ Selectable_ Buttons	out Amount	:Product_ Vending_ Controller	s_{11}

Figure 20-6. Relation $SSMR_{VM}$

Ψ_1	Δ					Ψ_2
	N	Ξ	Λ	Θ	Γ	
	g_{21}					
s_{21}	*CAL*	Customer	Return_ Payment_ Request		:Return_ Payment_ Button	s_{22}
	v_{22}					
s_{22}	*CAL*	:Return_ Payment_ Button	Return_ Payment		:Product_ Vending_ Controller	s_{23}
	v_{23}					
s_{23}	*CAL*	:Product_ Vending_ Controller	Return_ Coin	in Return_Amount	:Coin_ Store	s_{24}
	v_{24}					
s_{24}	*CAL*	:Coin_ Dispenser	Dispense_ Coin	out Coins	:Coin_ Store	s_{25}
	g_{25}					
s_{25}	*CAL*	Customer	Deliver_ Coin	out Coins	:Coin_ Dispenser	s_{26}
	v_{14}					
s_{26}	*CAL*	:Product_ Selection_ Buttons	Refresh_ Selectable_ Buttons	out Amount	:Product_ Vending_ Controller	s_{21}

Figure 20-6 (continued). Relation $SSMR_{\mathrm{VM}}$

‖

Ψ_1	N	Ξ	Λ	Θ	Γ	Ψ_2
			Δ			
			g_{31}			
s_{31}	*CAL*	Customer	Selection_ Request		:Product_ Selection_ Buttons	s_{32}
			v_{32}			
s_{32}	*CAL*	:Product Selection Buttons	Product_ Select	in Product_Number; Product_Value	:Product Vending Controller	s_{33}
			v_{33}			
s_{33}	*CAL*	:Product Vending Controller	Pick_ Product	in Product_Number	:Product_ Store	s_{34}
			v_{34}			
s_{34}	*CAL*	:Product_ Dispenser	Dispense_ Product	out Product	:Product_ Store	s_{35}
			g_{35}			
s_{35}	*CAL*	Customer	Deliver_ Product	out Product	:Product_ Dispenser	s_{36}
			v_{23}			
s_{36}	*CAL*	:Product_ Vending_ Controller	Return_ Coin	in Return_Amount	:Coin_ Store	s_{37}
			v_{24}			
s_{37}	*CAL*	:Coin_ Dispenser	Dispense_ Coin	out Coins	:Coin_ Store	s_{38}
			g_{25}			
s_{38}	*CAL*	Customer	Deliver_ Coin	out Coins	:Coin_ Dispenser	s_{39}
			v_{14}			
s_{39}	*CAL*	:Product_ Selection_ Buttons	Refresh_ Selectable_ Buttons	out Amount	:Product_ Vending_ Controller	s_{31}

Figure 20-6 (continued). Relation $SSMR_{\mathrm{VM}}$

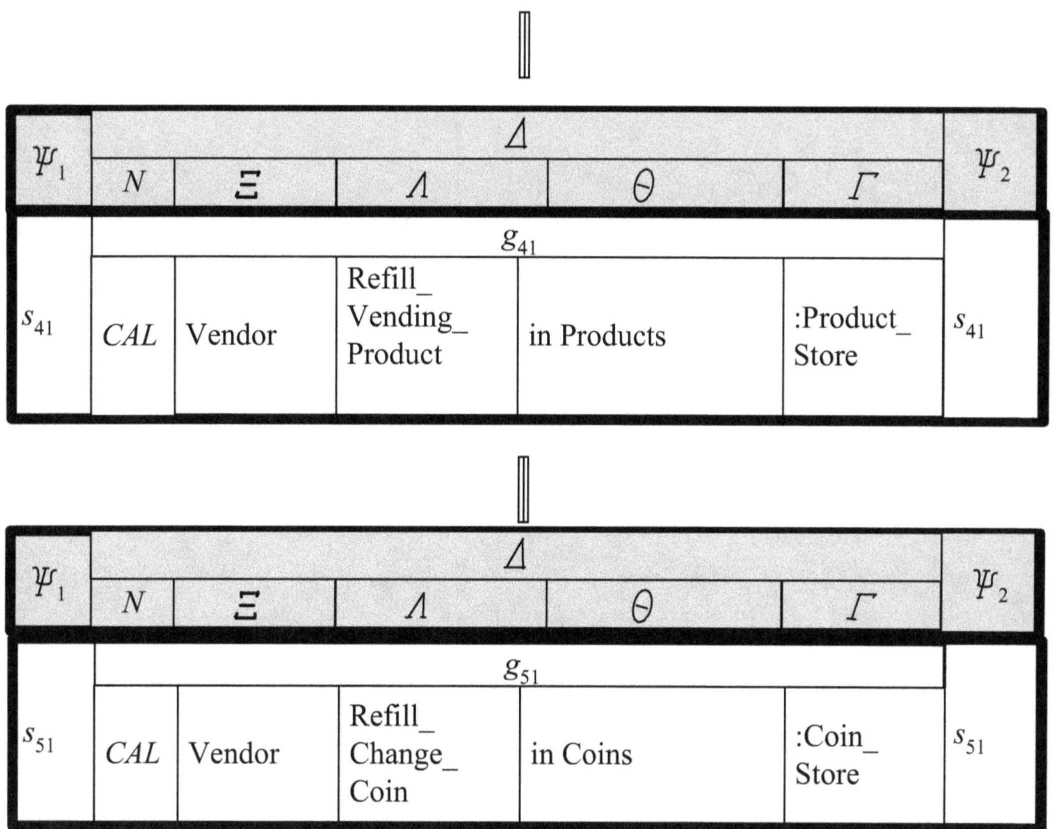

Ψ_1	Δ					Ψ_2
	N	Ξ	Λ	Θ	Γ	
	g_{41}					
s_{41}	*CAL*	Vendor	Refill_ Vending_ Product	in Products	:Product_ Store	s_{41}

Ψ_1	Δ					Ψ_2
	N	Ξ	Λ	Θ	Γ	
	g_{51}					
s_{51}	*CAL*	Vendor	Refill_ Change_ Coin	in Coins	:Coin_ Store	s_{51}

Figure 20-6 (continued). Relation $SSMR_{\text{VM}}$

Chapter 21: Projecting the SysML Use Case Diagram from the SBC State Machine of the Vending Machine

We use two steps to project the SysML use case diagram from the SBC state machine of the vending machine. First, we project the UCD relation from the state machine of the vending machine. Second, we will draw the SysML use case diagram from the UCD relation of the vending machine.

21-1 Projecting the UCD Relation from the SBC State Machine of the Vending Machine

The SSM relation $SSMR_{VM} \subseteq \Psi_1$ X N X Ξ X Λ X Θ X Γ X Ψ_2, defined as "$SSMR_1 \| SSMR_2 \| SSMR_3 \| SSMR_{24} \| SSMR_5$" and shown in Figure 20-6, is used to represent the SBC state machine SSM_{VM} of the vending machine.

We apply the algorithm of projecting the UCD relation (i.e., $UCDR_{VM}$) from the SSM relation (i.e., $SSMR_{VM}$) of the vending machine. After the projection, we get the relation $UCDR_{VM} \subseteq B$ X U as shown in Figure 21-1.

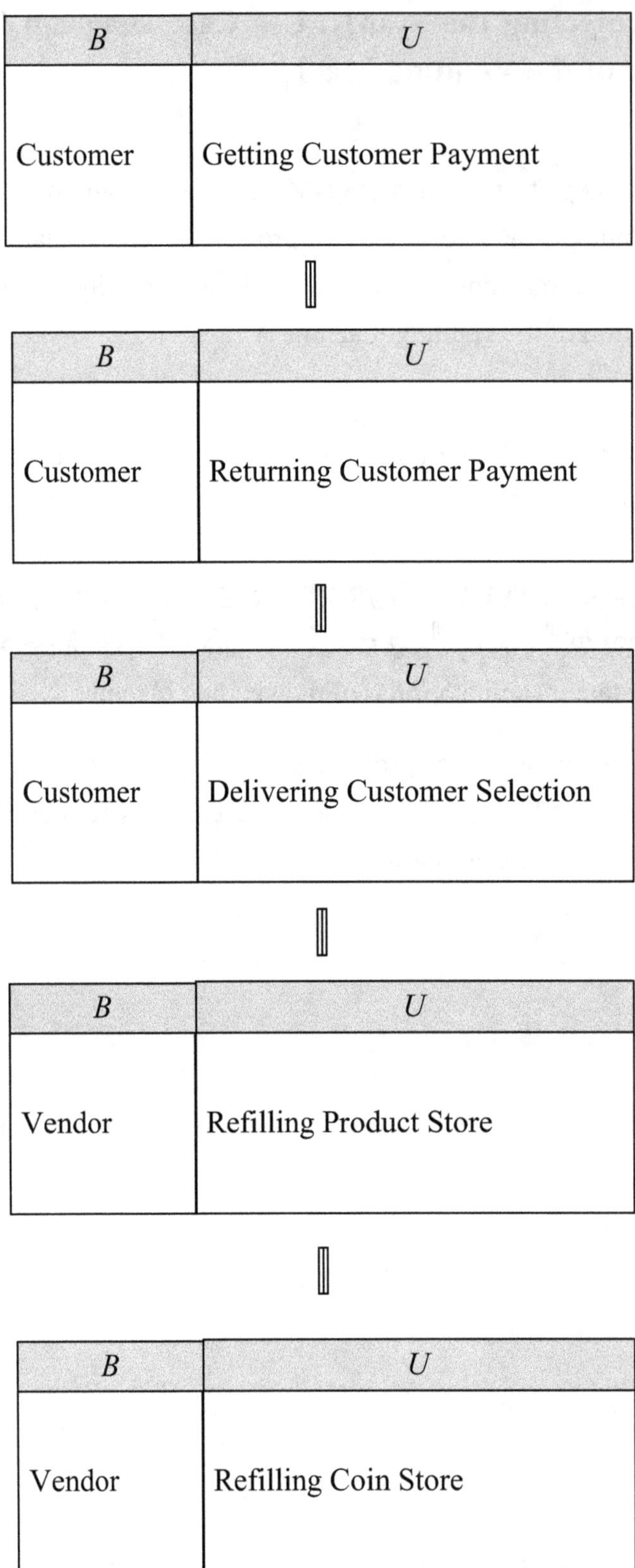

Figure 21-1. Relation $UCDR_{\mathrm{VM}}$

21-2 Achieving the Use Case Diagram from the UCD Relation of the Vending Machine

From the UCD relation $UCDR_{VM}$, we draw the corresponding SysML use case diagram of the vending machine, as shown in Figure 21-2.

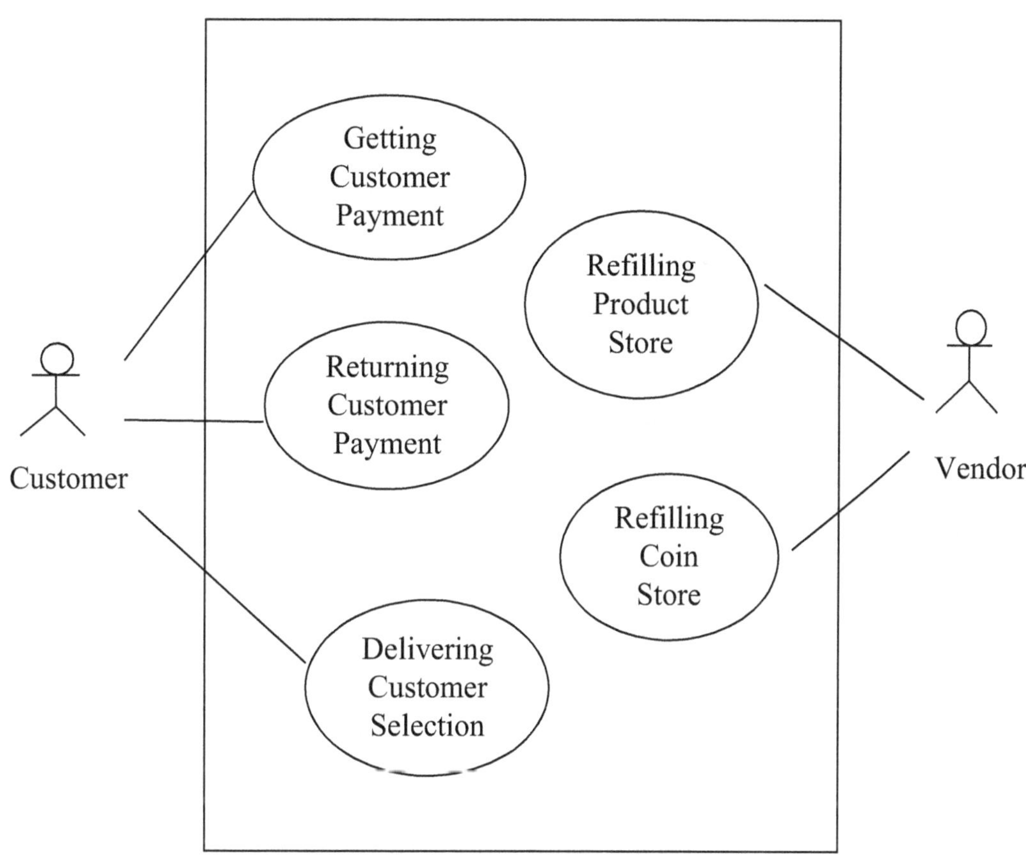

Figure 21-2. Use Case Diagram of the Vending Machine

Chapter 22: Projecting the SysML State Machine from the SBC State Machine of the Vending Machine

We use two steps to project a SysML state machine from the SBC state machine of the vending machine. First, we project the STM relation from the state machine of the vending machine. Second, we will draw the SysML state machine from the STM relation of the vending machine.

22-1 Projecting the STM Relation from the SBC State Machine of the Vending Machine

The SSM relation $SSMR_{VM} \subseteq \Psi_1 \text{ X } N \text{ X } \Xi \text{ X } \Lambda \text{ X } \Theta \text{ X } \Gamma \text{ X } \Psi_2$, defined as "$SSMR_1 \| SSMR_2 \| SSMR_3 \| SSMR_{24} \| SSMR_5$" and shown in Figure 20-6, is used to represent the SBC state machine SSM_{VM} of the vending machine.

We apply the algorithm of projecting the STM relation (i.e., $STMR_{VM}$) from the SSM relation (i.e., $SSMR_{VM}$) of the vending machine. After the projection, we get the relation $STMR_{VM} \subseteq \Psi_1 \text{ X } N \text{ X } \Lambda \text{ X } \Psi_2$ as shown in Figure 22-1.

Ψ_1	N	Λ	Ψ_2
s_{11}	*CAL*	Accept_ Coin	s_{12}
s_{12}	*CAL*	Deposit_ Coin	s_{13}
s_{13}	*CAL*	Accumulate_ Payment	s_{14}
s_{14}	*CAL*	Refresh_ Selectable_ Buttons	s_{11}

Figure 22-1. Relation $SMDR_{\mathrm{VM}}$

‖

Ψ_1	N	Λ	Ψ_2
s_{21}	*CAL*	Return_ Payment_ Request	s_{22}
s_{22}	*CAL*	Return_ Payment	s_{23}
s_{23}	*CAL*	Return_ Coin	s_{24}
s_{24}	*CAL*	Dispense_ Coin	s_{25}
s_{25}	*CAL*	Deliver_ Coin	s_{26}
s_{26}	*CAL*	Refresh_ Selectable_ Buttons	s_{21}

Figure 22-1 (continued). Relation $SMDR_{\mathrm{VM}}$

‖

Ψ_1	N	Λ	Ψ_2
s_{31}	*CAL*	Selection_ Request	s_{32}
s_{32}	*CAL*	Product_ Select	s_{33}
s_{33}	*CAL*	Pick_ Product	s_{34}
s_{34}	*CAL*	Dispense_ Product	s_{35}
s_{35}	*CAL*	Deliver_ Product	s_{36}
s_{36}	*CAL*	Return_ Coin	s_{37}
s_{37}	*CAL*	Dispense_ Coin	s_{38}
s_{38}	*CAL*	Deliver_ Coin	s_{39}
s_{39}	*CAL*	Refresh_ Selectable_ Buttons	s_{31}

Figure 22-1 (continued). Relation $SMDR_{VM}$

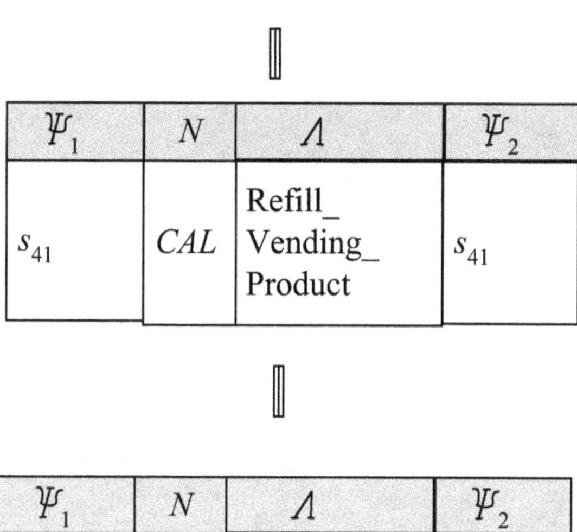

Ψ_1	N	Λ	Ψ_2
s_{41}	*CAL*	Refill_ Vending_ Product	s_{41}

Ψ_1	N	Λ	Ψ_2
s_{51}	*CAL*	Refill_ Change_ Coin	s_{51}

Figure 22-1 (continued). Relation $SMDR_{\mathrm{VM}}$

22-2 Achieving the SysML State Machine from the STM Relation of the Vending Machine

From the STM relation $STMR_{\mathrm{VM}}$, we draw the corresponding SysML state machine of the vending machine, as shown in Figure 22-2.

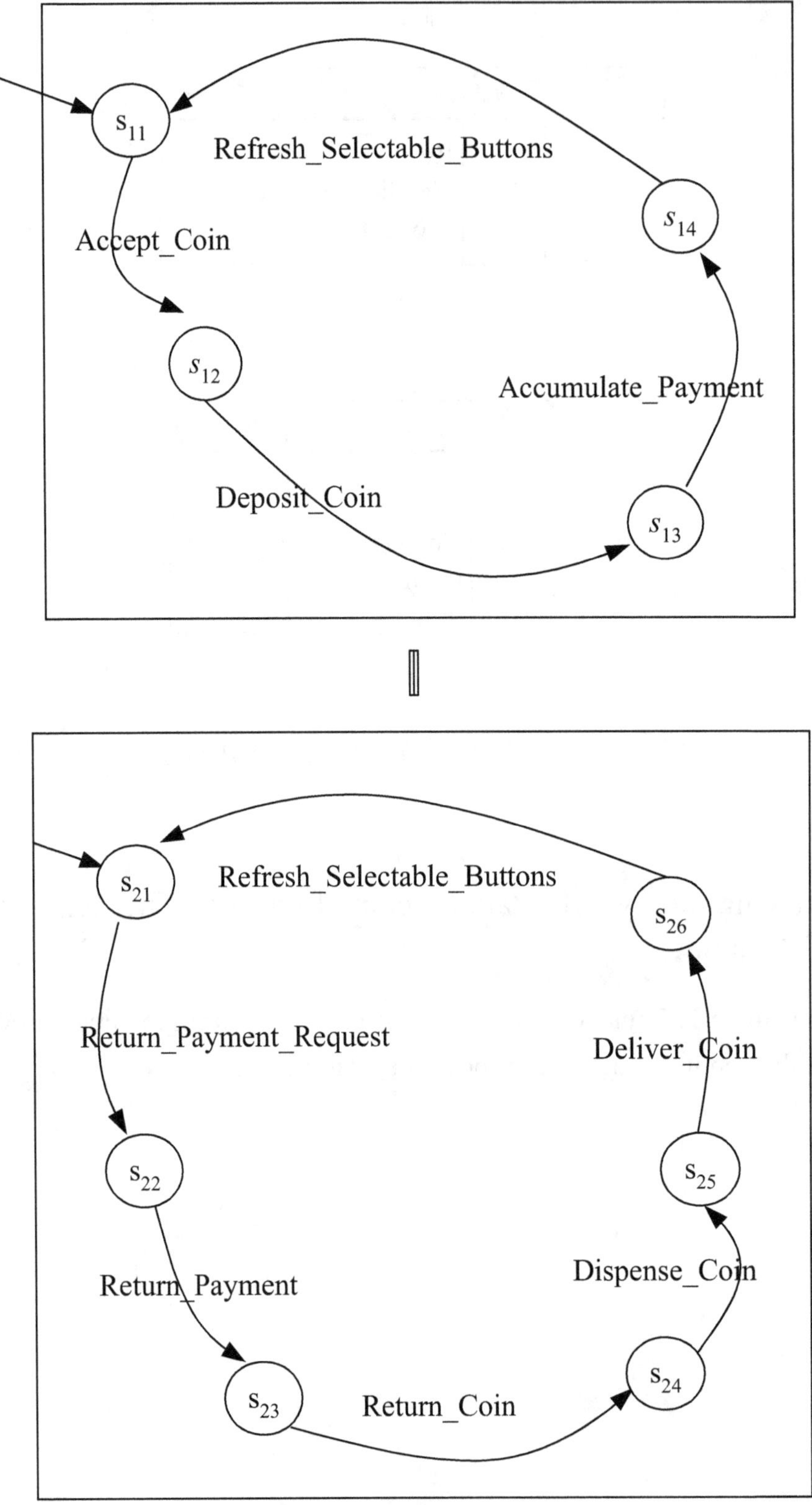

Figure 22-2. SysML State Machine Diagram of the Vending Machine

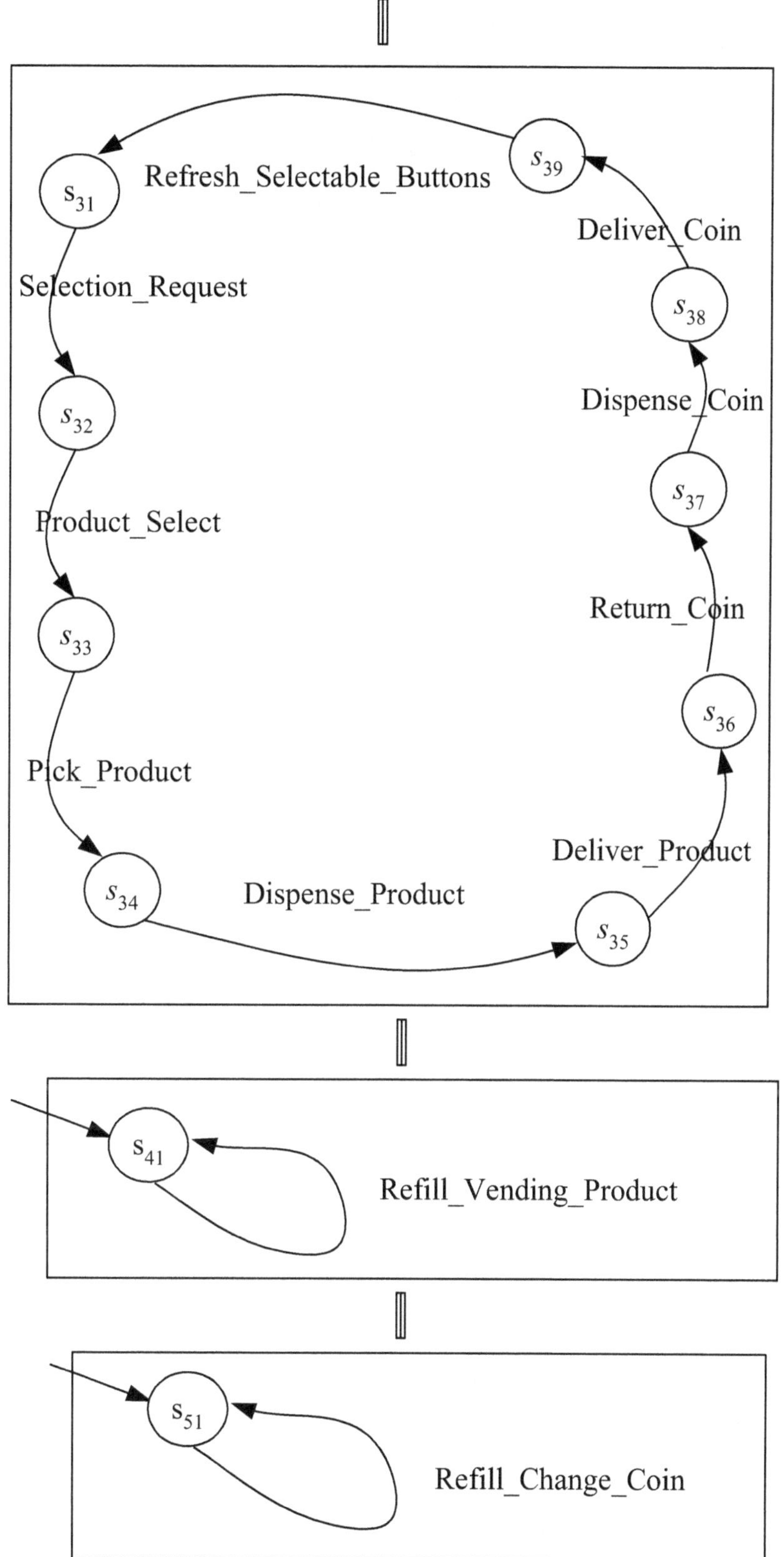

Figure 22-2 (continued). SysML State Machine Diagram of the Vending Machine

Chapter 23: Projecting the SysML Activity Diagram from the SBC State Machine of the Vending Machine

We use two steps to project a SysML activity diagram from the SBC state machine of the vending machine. First, we project the AD relation from the state machine of the Vending machine. Second, we will draw the SysML activity diagram from the AD relation of the vending machine.

23-1 Projecting the AD Relation from the SBC State Machine of the Vending Machine

The SSM relation $SSMR_{\mathrm{VM}} \subseteq \Psi_1 \times N \times \Xi \times \Lambda \times \Theta \times \Gamma \times \Psi_2$, defined as "$SSMR_1 \| SSMR_2 \| SSMR_3 \| SSMR_{24} \| SSMR_5$" and shown in Figure 20-4, is used to represent the SBC state machine SSM_{VM} of the vending machine.

We apply the algorithm of projecting the AD relation (i.e., ADR_{VM}) from the SSM relation (i.e., $SSMR_{\mathrm{VM}}$) of the vending machine. After the projection, we get the relation $ADR_{\mathrm{VM}} \subseteq \Psi_1 \times N \times \Lambda \times \Theta \times \Gamma \times \Psi_2$ as shown in Figure 23-1.

Ψ_1	N	Λ	Θ	Γ	Ψ_2
s_{11}	*CAL*	Accept_ Coin	in Coin	:Coin_ Receptacle	s_{12}
s_{12}	*CAL*	Deposit_ Coin	in Coin	:Coin_ Store	s_{13}
s_{13}	*CAL*	Accumulate_ Payment	in Coin_Value	:Product_ Vending_ Controller	s_{14}
s_{14}	*CAL*	Refresh_ Selectable_ Buttons	out Amount	:Product_ Vending_ Controller	s_{11}

Figure 23-1. Relation ADR_{VM}

‖

Ψ_1	N	Λ	Θ	Γ	Ψ_2
s_{21}	*CAL*	Return_ Payment_ Request		:Return_ Payment_ Button	s_{22}
s_{22}	*CAL*	Return_ Payment		:Product_ Vending_ Controller	s_{23}
s_{23}	*CAL*	Return_ Coin	in Order_Id; out Invoice	:Coin_ Store	s_{24}
s_{24}	*CAL*	Dispense_ Coin	in Credit_ Card_Id; in Amount; out Commit_ Response	:Coin_ Store	s_{25}
s_{25}	*CAL*	Deliver_ Coin	in Order_Id; in Amount; out Order_ Status	:Coin_ Dispenser	s_{26}
s_{26}	*CAL*	Refresh_ Selectable_ Buttons	out Order_ Status	:Product_ Vending_ Controller	s_{21}

Figure 23-1 (continued). Relation ADR_{VM}

$\|$

Ψ_1	N	Λ	Θ	Γ	Ψ_2
s_{31}	*CAL*	Selection_ Request		:Product_ Selection_ Buttons	s_{32}
s_{32}	*CAL*	Product_ Select	in Product_Number; Product_Value	:Product Vending Controller	s_{33}
s_{33}	*CAL*	Pick_ Product	in Product_Number	:Product_ Store	s_{34}
s_{34}	*CAL*	Dispense_ Product	out Product	:Product_ Store	s_{35}
s_{35}	*CAL*	Deliver_ Product	out Product	:Product_ Dispenser	s_{36}
s_{36}	*CAL*	Return_ Coin	in Return_Amount	:Coin_ Store	s_{37}
s_{37}	*CAL*	Dispense_ Coin	out Coins	:Coin_ Store	s_{38}
s_{38}	*CAL*	Deliver_ Coin	out Coins	:Coin_ Dispenser	s_{39}
s_{39}	*CAL*	Refresh_ Selectable_ Buttons	out Amount	:Product_ Vending_ Controller	s_{31}

Figure 23-1 (continued). Relation ADR_{VM}

$\|$

Ψ_1	N	Λ	Θ	Γ	Ψ_2
s_{41}	*CAL*	Refill_ Vending_ Product	in Products	:Product_ Store	s_{41}

$\|$

Ψ_1	N	Λ	Θ	Γ	Ψ_2
s_{51}	*CAL*	Refill_ Change_ Coin	in Coins	:Coin_ Store	s_{51}

Figure 23-1 (continued). Relation ADR_{VM}

23-2 Achieving the Activity Diagram from the AD Relation of the Vending Machine

From the AD relation ADR_{VM}, we draw the corresponding SysML activity diagram of the vending machine, as shown in Figure 23-2.

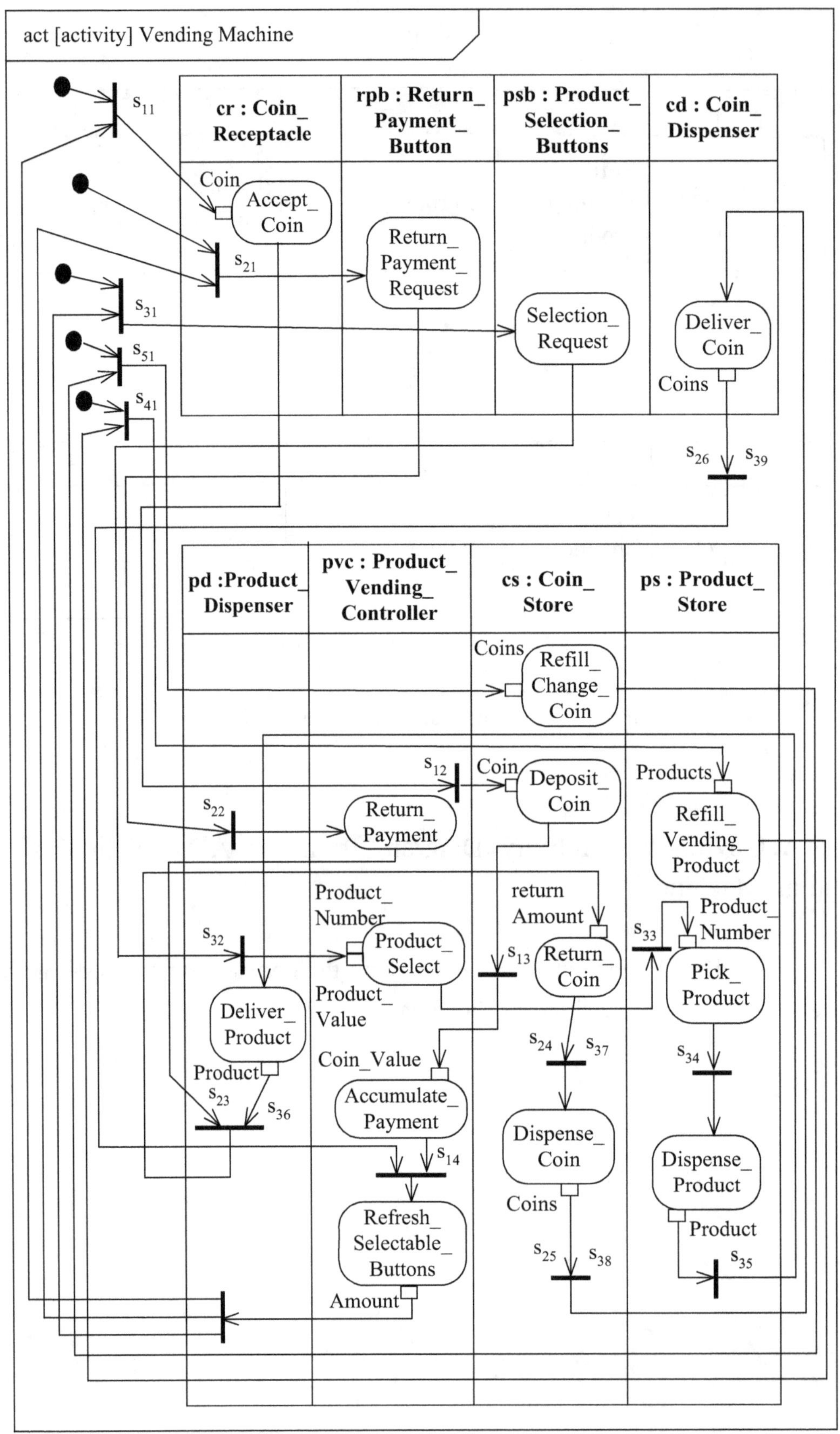

Figure 23-2. SysML Activity Diagram of the Vending Machine

Chapter 24: Projecting the SysML Sequence Diagram from the SBC State Machine of the Vending Machine

We use two steps to project a SysML sequence diagram from the SBC state machine of the vending machine. First, we project the SqD relation from the state machine of the Vending machine. Second, we will draw the SysML sequence diagram from the SqD relation of the vending machine.

24-1 Projecting the SqD Relation from the SBC State Machine of the Vending Machine

The SSM relation $SSMR_{VM} \subseteq \Psi_1 \text{ X } N \text{ X } \Xi \text{ X } \Lambda \text{ X } \Theta \text{ X } \Gamma \text{ X } \Psi_2$, defined as "$SSMR_1 \| SSMR_2 \| SSMR_3 \| SSMR_{24} \| SSMR_5$" and shown in Figure 20-4, is used to represent the SBC state machine SSM_{VM} of the vending machine.

We apply the algorithm of projecting the SqD relation (i.e., $SqDR_{VM}$) from the SSM relation (i.e., $SSMR_{VM}$) of the vending machine. After the projection, we get the relation $SqDR_{VM} \subseteq E \text{ X } N \text{ X } \Xi \text{ X } \Lambda \text{ X } \Theta \text{ X } \Gamma$ as shown in Figure 24-1.

E	*N*	*Ξ*	*Λ*	*Θ*	*Γ*
1	*CAL*	Customer	Accept_ Coin	in Coin	:Coin_ Receptacle
2	*CAL*	:Coin_ Receptacle	Deposit_ Coin	in Coin	:Coin_ Store
3	*CAL*	:Coin_ Receptacle	Accumulate_ Payment	in Coin_Value	:Product_ Vending_ Controller
4	*CAL*	:Product_ Selection_ Buttons	Refresh_ Selectable_ Buttons	out Amount	:Product_ Vending_ Controller

Figure 24-1. Relation $SqDR_{\mathrm{VM}}$

‖

E	N	Ξ	Λ	Θ	Γ
1	*CAL*	Customer	Return_Payment_Request		:Return_Payment_Button
2	*CAL*	:Return_Payment_Button	Return_Payment		:Product_Vending_Controller
3	*CAL*	:Product_Vending_Controller	Return_Coin	in Return_Amount	:Coin_Store
4	*CAL*	:Coin_Dispenser	Dispense_Coin	out Coins	:Coin_Store
5	*CAL*	Customer	Deliver_Coin	out Coins	:Coin_Dispenser
6	*CAL*	:Product_Selection_Buttons	Refresh_Selectable_Buttons	out Amount	:Product_Vending_Controller

Figure 24-1 (continued). Relation $SqDR_{VM}$

$\|$

E	N	Ξ	Λ	Θ	Γ
1	*CAL*	Customer	Selection_Request		:Product_Selection_Buttons
2	*CAL*	:Product_Selection_Buttons	Product_Select	in Product_Number; Product_Value	:Product_Vending_Controller
3	*CAL*	:Product_Vending_Controller	Pick_Product	in Product_Number	:Product_Store
4	*CAL*	:Product_Dispenser	Dispense_Product	out Product	:Product_Store
5	*CAL*	Customer	Deliver_Product	out Product	:Product_Dispenser
6	*CAL*	:Product_Vending_Controller	Return_Coin	in Return_Amount	:Coin_Store
7	*CAL*	:Coin_Dispenser	Dispense_Coin	out Coins	:Coin_Store
8	*CAL*	Customer	Deliver_Coin	out Coins	:Coin_Dispenser
9	*CAL*	:Product_Selection_Buttons	Refresh_Selectable_Buttons	out Amount	:Product_Vending_Controller

Figure 24-1 (continued). Relation $SqDR_{\mathrm{VM}}$

‖

E	N	Ξ	Λ	Θ	Γ
1	*CAL*	Vendor	Refill_ Vending_ Product	in Products	:Product_ Store

‖

E	N	Ξ	Λ	Θ	Γ
1	*CAL*	Vendor	Refill_ Change_ Coin	in Coins	:Coin_ Store

Figure 24-1 (continued). Relation $SqDR_{\mathrm{VM}}$

24-2 Achieving the Sequence Diagram from the SqD Relation of the Vending Machine

From the SqD relation $SqDR_{VM}$, we draw the corresponding SysML sequence diagram of the vending machine, as shown in Figure 24-2.

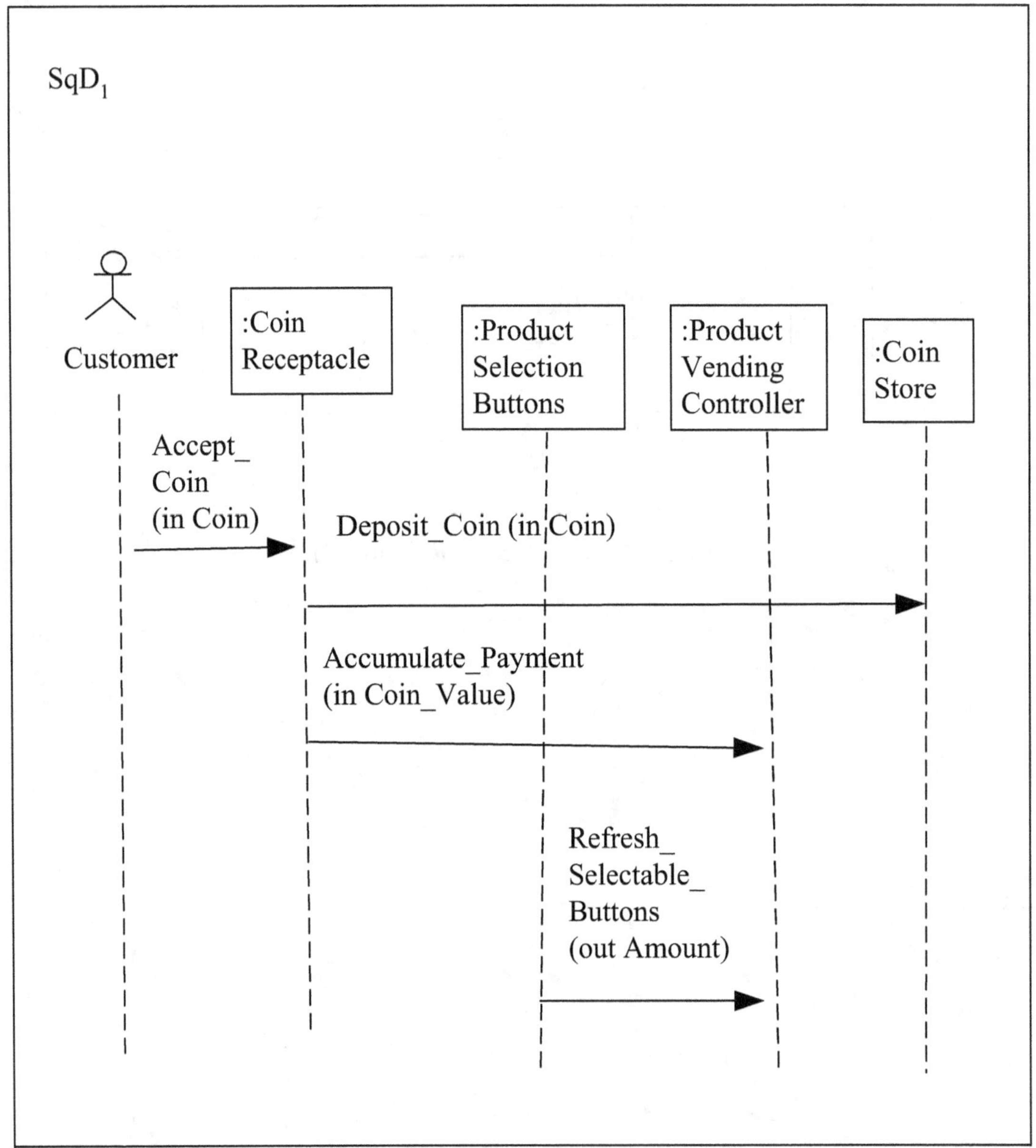

Figure 24-2. Sequence Diagram of the Vending Machine

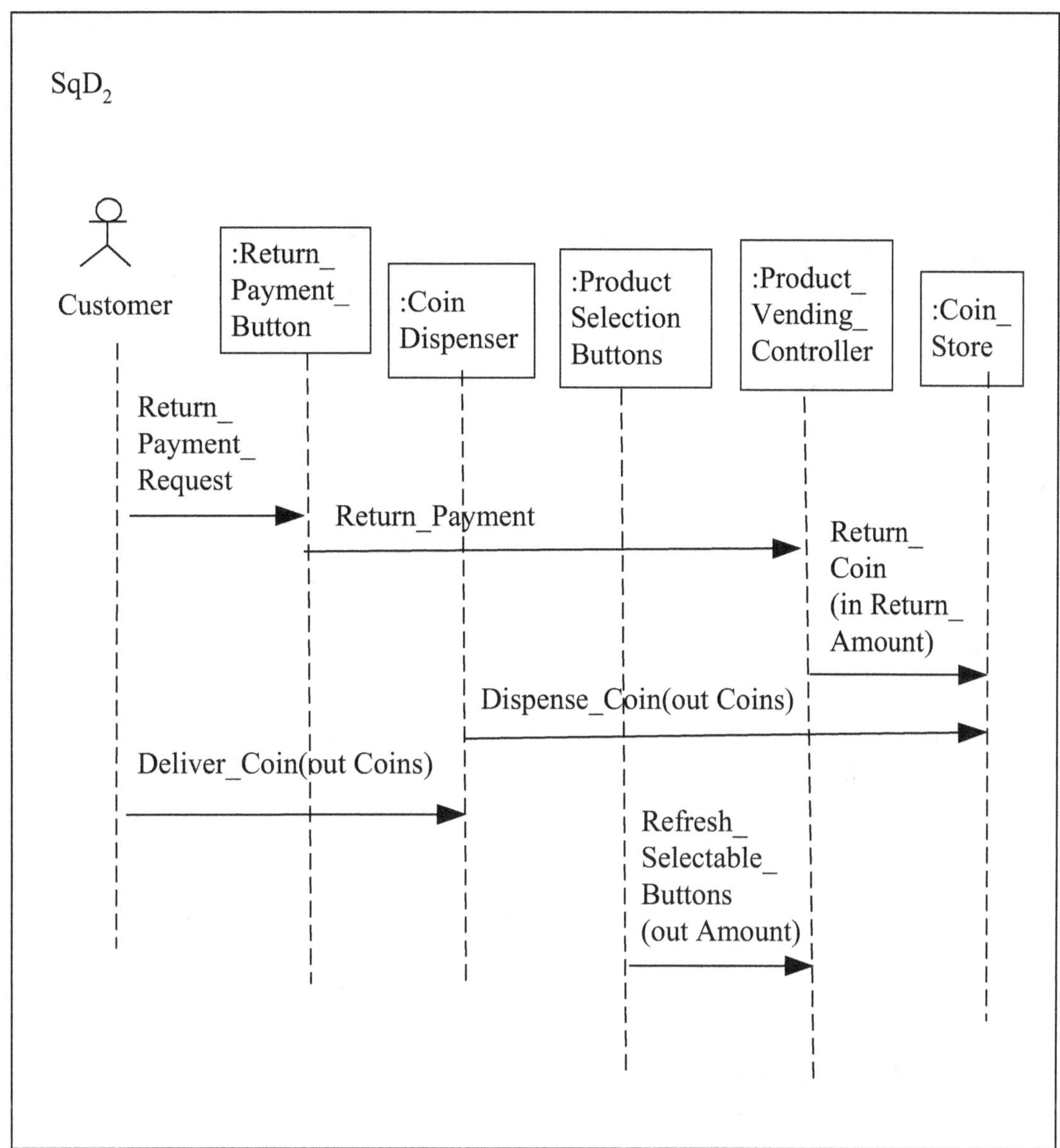

Figure 24-2 (continued). Sequence Diagram of the Vending Machine

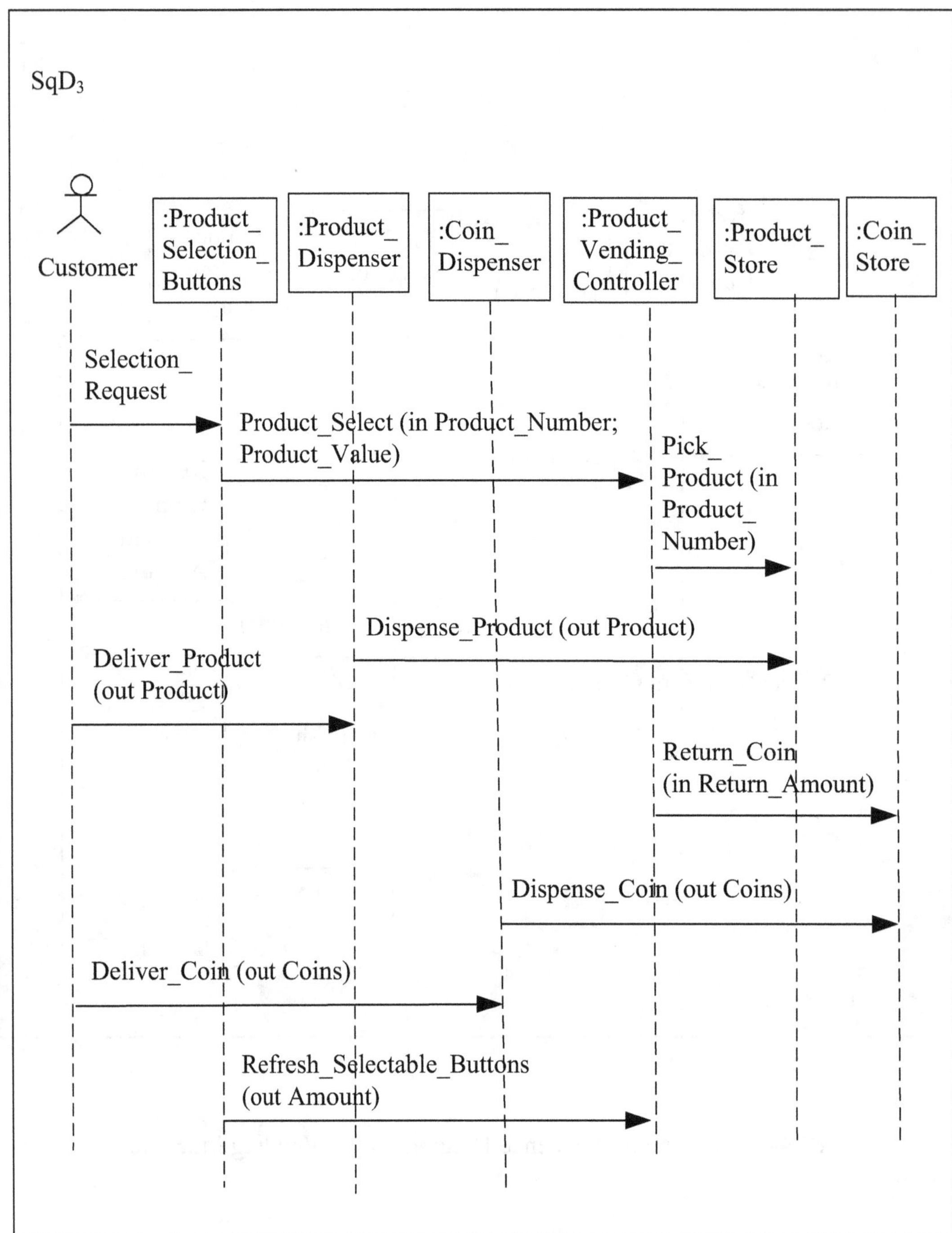

Figure 24-2 (continued). Sequence Diagram of the Vending Machine

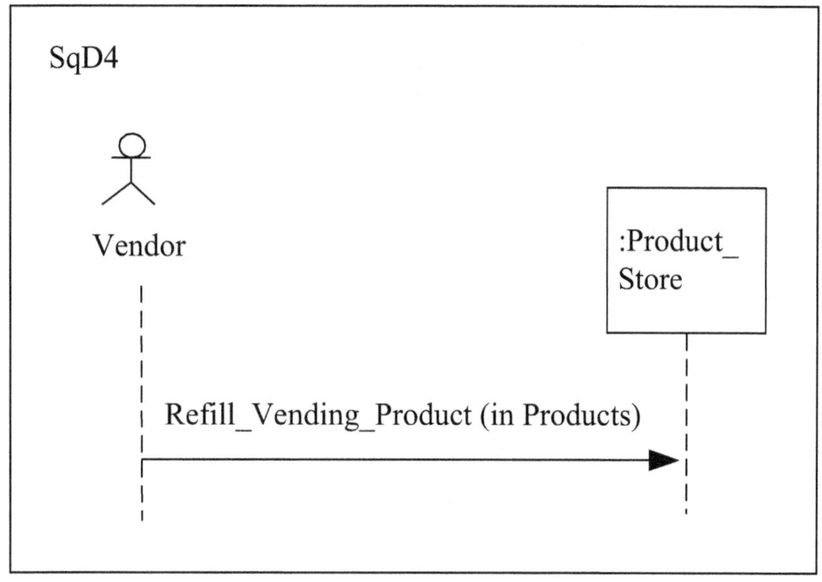

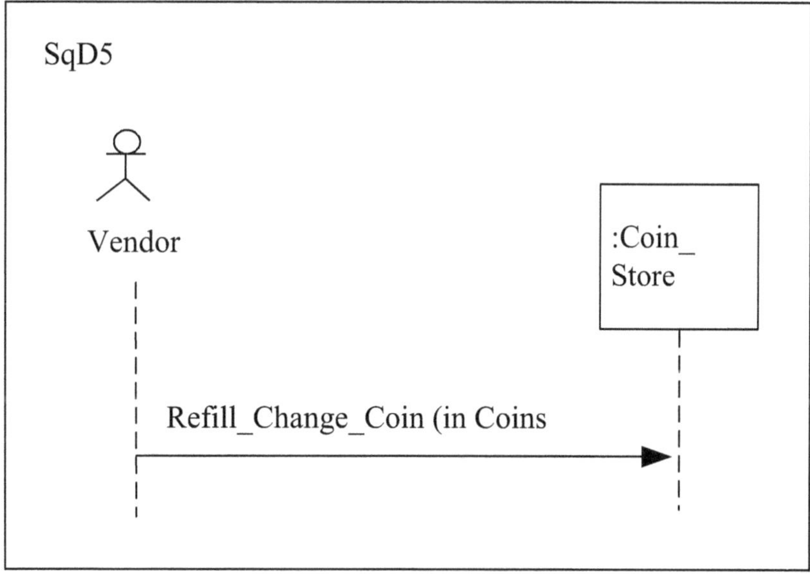

Figure 24-2 (continued). Sequence Diagram of the Vending Machine

Chapter 25: Projecting the SysML Internal Block Diagram from the SBC State Machine of the Vending Machine

We use two steps to project the SysML internal block diagram from the SBC state machine of the vending machine. First, we project the IBD relation from the state machine of the Vending machine. Second, we will draw the SysML internal block diagram from the IBD relation of the vending machine.

25-1 Projecting the IBD Relation from the SBC State Machine of the Vending Machine

The SSM relation $SSMR_{\mathrm{VM}} \subseteq \Psi_1 \mathrm{X}\ N\ \mathrm{X}\ \Xi\ \mathrm{X}\ \Lambda\ \mathrm{X}\ \Theta\ \mathrm{X}\ \Gamma\ \mathrm{X}\ \Psi_2$, defined as "$SSMR_1 \| SSMR_2 \| SSMR_3 \| SSMR_{24} \| SSMR_5$" and shown in Figure 20-4, is used to represent the SBC state machine SSM_{VM} of the vending machine.

We apply the algorithm of projecting the IBD relation (i.e., $IBDR_{\mathrm{VM}}$) from the SSM relation (i.e., $SSMR_{\mathrm{VM}}$) of the vending machine. After the projection, we get the relation $IBDR_{\mathrm{VM}} \subseteq N\ \mathrm{X}\ \Xi\ \mathrm{X}\ \Lambda\ \mathrm{X}\ \Theta\ \mathrm{X}\ \Gamma$ as shown in Figure 25-1.

N	Ξ	Λ	Θ	Γ
CAL	Customer	Accept_ Coin	in Coin	:Coin_ Receptacle
CAL	:Coin_ Receptacle	Deposit_ Coin	in Coin	:Coin_ Store
CAL	:Coin_ Receptacle	Accumulate_ Payment	in Coin_Value	:Product_ Vending_ Controller
CAL	:Product_ Selection_ Buttons	Refresh_ Selectable_ Buttons	out Amount	:Product_ Vending_ Controller

Figure 25-1. Relation $IBDR_{\mathrm{VM}}$

‖

N	Ξ	Λ	Θ	Γ
CAL	Customer	Return_ Payment_ Request		:Return_ Payment_ Button
CAL	:Return_ Payment_ Button	Return_ Payment		:Product_ Vending_ Controller
CAL	:Product_ Vending_ Controller	Return_ Coin	in Return_ Amount	:Coin_ Store
CAL	:Coin_ Dispenser	Dispense_ Coin	out Coins	:Coin_ Store
CAL	Customer	Deliver_ Coin	out Coins	:Coin_ Dispenser
CAL	:Product_ Selection_ Buttons	Refresh_ Selectable_ Buttons	out Amount	:Product_ Vending_ Controller

Figure 25-1 (continued). Relation $IBDR_{\mathrm{VM}}$

N	Ξ	Λ	Θ	Γ
CAL	Customer	Selection_Request		:Product_Selection_Buttons
CAL	:Product_Selection_Buttons	Product_Select	in Product_Number; Product_Value	:Product_Vending_Controller
CAL	:Product_Vending_Controller	Pick_Product	in Product_Number	:Product_Store
CAL	:Product_Dispenser	Dispense_Product	out Product	:Product_Store
CAL	Customer	Deliver_Product	out Product	:Product_Dispenser
CAL	:Product_Vending_Controller	Return_Coin	in Return_Amount	:Coin_Store
CAL	:Coin_Dispenser	Dispense_Coin	out Coins	:Coin_Store
CAL	Customer	Deliver_Coin	out Coins	:Coin_Dispenser
CAL	:Product_Selection_Buttons	Refresh_Selectable_Buttons	out Amount	:Product_Vending_Controller

Figure 25-1 (continued). Relation $IBDR_{\mathrm{VM}}$

$\|$

N	Ξ	Λ	Θ	Γ
CAL	Vendor	Refill_ Vending_ Product	in Products	:Product_ Store

$\|$

N	Ξ	Λ	Θ	Γ
CAL	Vendor	Refill_ Change_ Coin	in Coins	:Coin_ Store

Figure 25-1 (continued). Relation $IBDR_{\mathrm{VM}}$

25-2 Achieving the Internal Block Diagram from the IBD Relation of the Vending Machine

From the IBD relation $IBDR_{\mathrm{VM}}$, we draw the corresponding SysML internal block diagram of the vending machine, as shown in Figure 25-2.

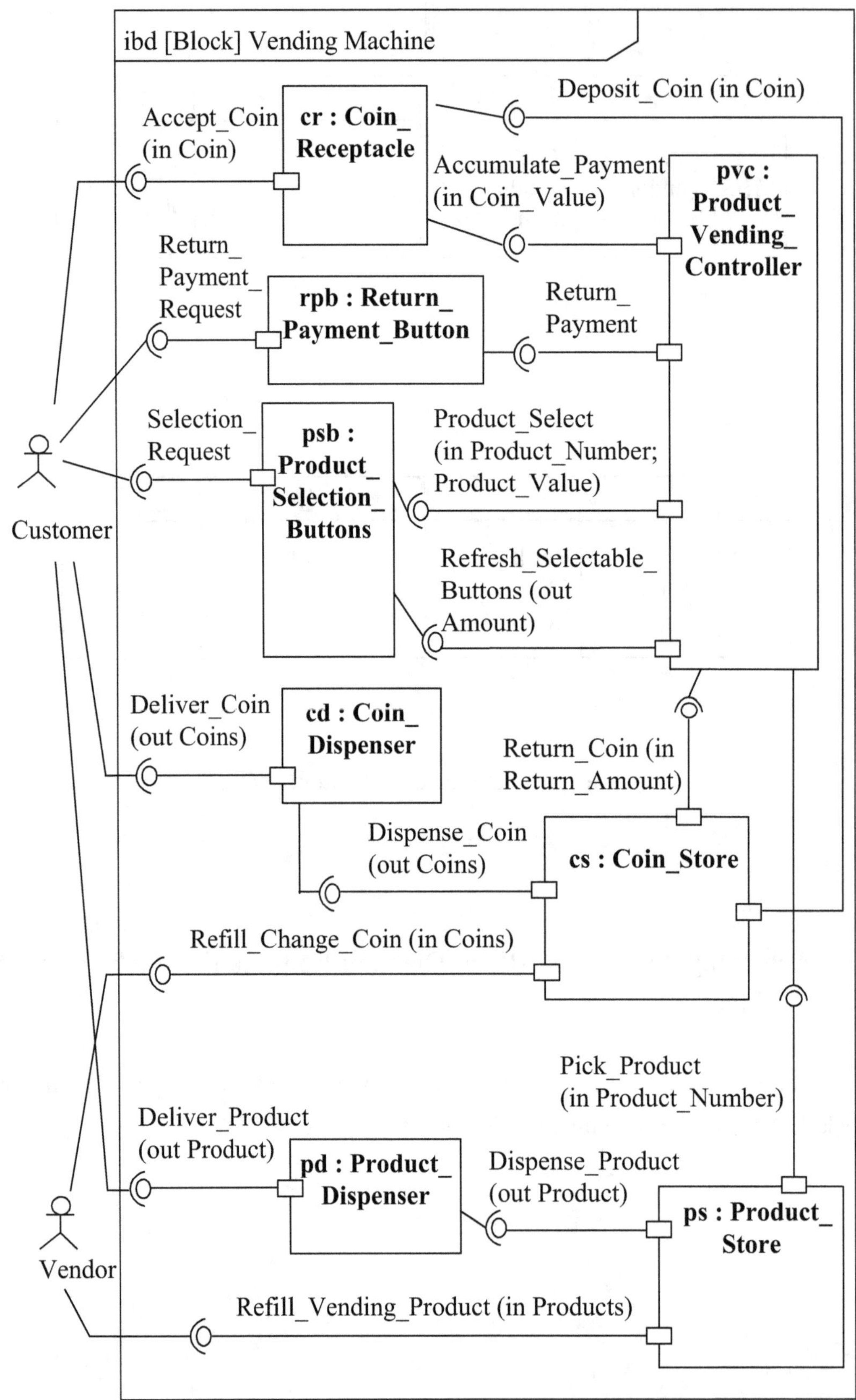

Figure 25-2. Internal Block Diagram of the Vending Machine

APPENDIX: ENTITY SET AND ENTITY NAME OF SBC STATE MACHINE

Entity set or relation	Entity name	Type of entity
K	$k_1, k_2...$	channel signatures
L	$l_1, l_2...$	operation call or operation return signatures
Λ	ch_1, ch_2	channel names
	op_1, op_2	operation names
Θ	$p_1, p_2...$	parameter lists
G	$g_1, g_2...$	type 1 interactions
V	$v_1, v_2...$	type 2 interactions
Δ	$a_1, a_2...$	type 1 or 2 interactions
	λ	internal interaction
	NOI	non-operable Interaction
Ω	$\propto_1, \propto_2...$	type_1_or_2 or internal or dummy interactions
f	$f_1, f_2...$	renaming functions
C	$c_1, c_2...$	guard conditions
R	$r_1, r_2...$	prefixes
Π	$\pi_1, \pi_2...$	code snippets

Entity set or relation	Entity name	Type of entity
	I, J,...	indexing sets
Ψ	$s_1, s_2...$	state expressions
X	$X_1, X_2...$	state variables
Φ	$A_1, A_2...$	state constants
B	$\beta_1, \beta_2...$	actors
Γ	$b_1, b_2...$	components
Ξ	$\rho_1, \rho_2...$	actors or components
N	$n_1, n_2...$	operation call or operation return tag

BIBLIOGRAPHY

[Alla15] Allaki, D. et al., "A New Taxonomy of Inconsistencies in UML Models with their Detection Methods for better MDE", *International Journal of Computer Science and Applications*, 12(1), pp. 48-65, 2015.

[Bash16] R. S. Bashir, R. S. et al., "UML Models Consistency Management: Guidelines for Software Quality Manager", *International Journal of Information Management*, 2016.

[Berg87] Bergstra, J. A. et al., "ACPτ: A Universal Axiom System for Process Specification," *CWI Quarterly* 15, 1987, pp. 3-23.

[Burd10] Burd, S. D., *Systems Architecture*, 6th Edition, Cengage Learning, 2010.

[Chao14a] Chao, W. S., *Systems Thingking 2.0: Architectural Thinking Using the SBC Architecture Description Language*, CreateSpace Independent Publishing Platform, 2014.

[Chao14b] Chao, W. S., *General Systems Theory 2.0: General Architectural Theory Using the SBC Architecture*, CreateSpace Independent Publishing Platform, 2014.

[Chao14c] Chao, W. S., *Software Modeling and Architecting: Structure-Behavior Coalescence for Software Architecture*, CreateSpace Independent Publishing Platform, 2014.

[Chao15a] Chao, W. S., *A Process Algebra For Systems Architecture: The Structure-Behavior Coalescence Approach*, CreateSpace Independent Publishing Platform, 2015.

[Chao15b] Chao, W. S., *An Observation Congruence Model For Systems Architecture: The Structure-Behavior Coalescence Approach*, CreateSpace Independent

Publishing Platform, 2015.

[Chao16] Chao, W. S., *System: Contemporary Concept, Definition, and Language*, CreateSpace Independent Publishing Platform, 2016.

[Chao17a] Chao, W. S., *Channel-Based Single-Queue SBC Process Algebra For Systems Definition: General Architectural Theory at Work*, CreateSpace Independent Publishing Platform, 2017.

[Chao17b] Chao, W. S., *Channel-Based Multi-Queue SBC Process Algebra For Systems Definition: General Architectural Theory at Work*, CreateSpace Independent Publishing Platform, 2017.

[Chao17c] Chao, W. S., *Channel-Based Infinite-Queue SBC Process Algebra For Systems Definition: General Architectural Theory at Work*, CreateSpace Independent Publishing Platform, 2017.

[Chao17d] Chao, W. S., *Operation-Based Single-Queue SBC Process Algebra For Systems Definition: General Architectural Theory at Work*, CreateSpace Independent Publishing Platform, 2017.

[Chao17e] Chao, W. S., *Operation-Based Multi-Queue SBC Process Algebra For Systems Definition: Unification of Systems Structure and Systems Behavior*, CreateSpace Independent Publishing Platform, 2017.

[Chao17f] Chao, W. S., *Operation-Based Infinite-Queue SBC Process Algebra For Systems Definition: Unification of Systems Structure and Systems Behavior*, CreateSpace Independent Publishing Platform, 2017.

[Chec99] Checkland, P., *Systems Thinking, Systems Practice: Includes a 30-Year Retrospective*, 1st Edition, Wiley, 1999.

[Craw15] Crawley, P. et al., *System Architecture: Strategy and Product Development for Complex Systems*, Prentice Hall, 2015.

[Dam06] Dam, S., *DoD Architecture Framework: A Guide to Applying System*

Engineering to Develop Integrated Executable Architectures, BookSurge Publishing, 2006.

[Date03] Date, C. J., *An Introduction to Database Systems*, 8th Edition, Addison Wesley, 2003.

[Dell13] Delligatti, L., SysML Distilled: A Brief Guide to the Systems Modeling Language, 1st Edition, Addison-Wesley, 2013.

[Denn08] Dennis, A. et al., *Systems Analysis and Design*, 4th Edition, Wiley, 2008.

[Dori95] Dori, D., "Object-Process Analysis: Maintaining the Balance between System Structure and Behavior," *Journal of Logic and Computation* 5(2), pp.227-249, 1995.

[Dori02] Dori, D., *Object-Process Methodology: A Holistic Systems Paradigm*, Springer Verlag, New York, 2002.

[Dori16] Dori, D., *Model-Based Systems Engineering with OPM and SysML*, Springer Verlag, New York, 2016.

[Enge02] Engels, G.. et al., "Consistency-preserving Model Evolution Through Transformations," *Proc. Int'l Conf. UML 2002*, pp. 212–227, 2002.

[Frie14] Friedenthal S. and A. Moore and R. Steiner, *A Practical Guide to SysML: The Systems Modeling Language*, 3rd Edition, Morgan Kaufmann; 2014.

[Hoar85] Hoare, C. A. R., *Communicating Sequential Processes*, Prentice-Hall, 1985.

[INCO04] INCOSE, Systems Engineering Vision, Int. Council Syst. Eng., 2004.

[Lale08] Laleau, R. et al., "Using Formal Metamodels to Check Consistency of Functional Views in Information Systems Specification," *Information & Software Technology*, pp. 797-814, 2008.

[Lin19] Lin, K. et al., "The Structure-Behavior Coalescence Approach for Systems Modeling," *IEEE Access*, Vol. 7, pp. 8609-8620, 2019.

[Malg06] Malgouyres, H. et al., "A UML Model Consistency Verification Approach

Based on Meta-modeling Formalization", *Proceedings of the 2006 ACM Symposium on Applied Computing*, pp. 1804-1809, 2006.

[Maie09] Maier, M. W., *The Art of Systems Architecting*, 3rd Edition, CRC Press, 2009.

[Miln89] Milner, R., *Communication and Concurrency*, Prentice-Hall, 1989.

[Miln99] Milner, R., *Communicating and Mobile Systems: the π-Calculus*, 1st Edition, Cambridge University Press, 1999.

[OMG 13a] OMG, Semantics of a Foundational Subset for Executable UML Models (fUML). *Object Management Group*, Needham, MA, 2013.

[OMG 13b] OMG, Action Language for Foundational UML (Alf). *Object Management Group*, Needham, MA, 2013.

[O'Rou03] O'Rourke, C. et al, *Enterprise Architecture Using the Zachman Framework*, 1st Edition, Course Technology, 2003.

[Pele00] Peleg, M. et al., "The Model Multiplicity Problem: Experimenting with Real-Time Specification Methods", *IEEE Tran. on Software Engineering*. 26 (8), pp. 742–759, 2000.

[Przi16] Przigoda, N. et al., "Analyzing Inconsistencies in UML/OCL Models", *Journal of Circuits, Systems and Computers*, 25(3), 2016.

[Quei83] Queille, J. P. et al., "Fairness and Related Properties in Transition Systems – A Temporal Logic to Deal with Fairness", *Acta Informatica*, Vol. 19: pp. 195-220, 1983.

[Rayn09] Raynard, B., *TOGAF The Open Group Architecture Framework 100 Success Secrets*, Emereo Pty Ltd, 2009.

[Roza11] Rozanski, N. et al., *Software Systems Architecture: Working With Stakeholders Using Viewpoints and Perspectives*, 2nd Edition, Addison-Wesley Professional, 2011.

[Sang03] Sangiorgi, D. et al., *The Pi-Calculus: A Theory of Mobile Processes*, Cambridge University Press, 2003.

[Weil08] Weilkiens, T., *Systems Engineering with SysML/UML: Modeling, Analysis, Design*, Morgan Kaufmann, 2008.

INDEX

I

L

M

N

O

P

R

S

T

U

www.ingramcontent.com/pod-product-compliance
Lightning Source LLC
Chambersburg PA
CBHW080825220526
45467CB00008B/2196

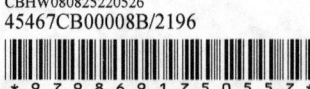

* 9 7 9 8 6 9 1 7 5 0 5 5 7 *